Studies in Fuzziness and Soft Computing

Volume 427

Series Editor

Janusz Kacprzyk, Systems Research Institute, Polish Academy of Sciences, Warsaw, Poland

The series "Studies in Fuzziness and Soft Computing" contains publications on various topics in the area of soft computing, which include fuzzy sets, rough sets, neural networks, evolutionary computation, probabilistic and evidential reasoning, multi-valued logic, and related fields. The publications within "Studies in Fuzziness and Soft Computing" are primarily monographs and edited volumes. They cover significant recent developments in the field, both of a foundational and applicable character. An important feature of the series is its short publication time and world-wide distribution. This permits a rapid and broad dissemination of research results.

Indexed by SCOPUS, DBLP, WTI Frankfurt eG, zbMATH, SCImago.

All books published in the series are submitted for consideration in Web of Science.

László T. Kóczy

Fuzzy Cognitive Maps

A Tool for the Modeling and Simulation of Processes and Systems

Springer

László T. Kóczy
Department of Information Technology
Széchenyi István University
Győr, Hungary

ISSN 1434-9922 ISSN 1860-0808 (electronic)
Studies in Fuzziness and Soft Computing
ISBN 978-3-031-37958-1 ISBN 978-3-031-37959-8 (eBook)
https://doi.org/10.1007/978-3-031-37959-8

This Springer imprint is published by the registered company Springer Nature Switzerland AG
The registered company address is: Gewerbestrasse 11, 6330 Cham, Switzerland

Foreword

Cognitive maps and fuzzy cognitive maps have been an area of intensive and productive research. As a modeling environment, they come with a number of appealing features. Knowledge representation, computing and optimization capabilities, and interpretability are the visible outstanding features of these models manifested vividly in numerous applications. In a nutshell, fuzzy cognitive maps are graph models augmented with computing realized at each node of the graph. We have been witnessing a wealth of learning schemes leading to new avenues of design of the maps. As nonlinear models, the maps require prudent analysis when it comes to a comprehensive analysis of the dynamics of their architectures and a thorough identification of dynamic patterns and their stability analysis.

This research monograph brings forward a number of timely and important aspects central to the area of intelligent systems. It covers a breadth of topics of cognitive maps ranging from systematic analysis, design, and applications. The book consists of ten chapters and provides the reader with a well-structured exposure of the subject matter. It covers the fundamentals and associated challenges, and brings formal ways of analyzing the maps, shows design approaches including those involving population-based optimization, demonstrates ways of coping with uncertainty (uncertain cognitive maps), brings interesting applications to a wealth of real-world problems such as those encountered in waste management.

The volume offers a systematic and coherent exposure of the material, underlines interesting innovative research insights, and positions the studies in the framework of timely applications.

There are a number of outstanding features that a reader can fully appreciate: (i) well-delineated relationships among concepts and real-world problems, (ii) a formation of essential linkages demonstrating ways of mapping a problem to a cognitive map, (iii) solid analysis, and (iv) involvement of advanced optimization methods with a demonstration of ways of their efficient usage in the construction of cognitive maps.

Professor László T. Kóczy is a truly outstanding leader in Computational Intelligence (CI) with a wealth of far-reaching, highly acclaimed, and impactful results in the analysis and synthesis of intelligent systems. This book is a testimony to his remarkable contributions that are instrumental to fostering further progress in the discipline of CI.

In sum, this a very much desirable book contributing to the body of knowledge of cognitive maps. There is no doubt that the reader will enjoy the timely and important publication.

Edmonton, Canada
January 2023

Witold Pedrycz

Preface

Cognitive Maps, and later also Fuzzy Cognitive Maps (FCM) have been in the focus of interest for several decades now. Although the terminology used in the latter refers to the fuzzy aspects of the method, the concept may be considered as part of the tool kit of the connectionist approaches. Thus, FCM is a kind of neural network, even though it can be described at the same time as a fuzzy graph. When considering FCM as a fuzzy graph, however, it should be clarified that it is directed and bipolar (the latter meaning that membership degrees are assigned an additional sign, thus each node may have an exciting or inhibiting effect on any other. These signed membership degrees thus express the positive or negative sense of the effect of the source node (concept) on the destination node (concept) of the directed edge. Nevertheless, the results in the literature analyzing the behavior of bipolar fuzzy graphs would lead to a path going away from the topic, not fitting into the goal of this book, and thus, the mathematical analysis of bipolar fuzzy graphs will be completely omitted here. It should be nevertheless noted that, in this representation, the set of directed edges $E = \{e_{ij}\}$ is mapped to the closed interval thus forming the set of all (directed) membership values M assigned to the edges

$$M = \{\mu_{ij}\} : E \to [-1, 1] \tag{1}$$

The fact that no detailed mathematical analysis is given here does not mean that we entirely neglect the necessity of an introductory presentation and analysis of FCM graphs. A brief chapter at the beginning will address and summarize these issues. There are plenty of monographs and textbooks that deal with various types FCMs from a mathematical point of view, and we did not feel the necessity to repeat the knowledge in these publications. The focus of this book lies on a different point. In order to clarify our intention, let us thus have a look at how this book was concepted and what its actual purpose is, according to the intentions of the contributors.

First of all, let us clarify that the book may be considered primarily as a monograph, rather so, than a kind of textbook, despite the fact that it was concepted partly based on some graduate (Ph.D.) research—at least, as the starting idea. As mentioned above, there is plenty of literature available on the FCMs; thus, we thought it would be redundant and unnecessary to summarize the detailed knowledge needed for a researcher to carry out the work of constructing and operating FCM models for particular applications. Nevertheless, we have included an introductory chapter in order to help recall the fundamental knowledge needed for understanding the contents, the "story" of this book, and we have referred to some of the related literature.

A story? It sounds rather exotic in the Preface of a scientific book that is supposed to be a kind of a monograph on a modeling approach. Nevertheless, there is a story here, which tells how and why the research described in this volume went, essentially starting with and based on two real-life case studies, stringing the chapters into a more or less continuous story presenting how the phases of this research were induced one by one, strung together like beads of a necklace. The "beads" are connected and building always on the previous "beads". They were starting by a real-life engineering type problem that had no satisfactory solution at the time, even, there was no sufficient model and explanation of the behavior of the system that could be found in the literature—at least, until we introduced the particular FCM approach for modeling it.

Sometime later, one more real-life problem was presented to our group, this time a financial management modeling problem, and this second problem, and our attempts to tackle it largely contributed to the development of several new models and some new mathematical results on FCM and also, to extended FCM approaches, adding to the story, continuing previous work. This other, similarly rather interesting practical problem was raised in the frame of a collaboration with a French university (École Central), which latter had been contracted by a French bank for carrying out a type of modeling work. It turned out, however, soon that the necessary tools were not available for the primary research group, and this way, our contribution was requested. Initiated by this new challenge, we did further connected research in order to clear some—rather important—aspects of FCM model stability. This problem started another string of beads connected somewhat loser to the first string. In summary, the content of this book is a two-ended "story" based on two case studies, resulting in some new recognitions in engineering, computer science, and even mathematics.

The somewhat unusual structure of the volume was also motivated because I thought, telling such a story based on two real-life research case studies, besides the actual results, may be used for interested Ph.D. students and, even postdoctoral researchers as a methodological example, as well. Thus the book has a secondary intention: to give a couple of examples of how a set research goal can be reached step by step, starting from premises collected from the available literature and the expert domain knowledge related to the application problem, and coming to conclusions with novel scientific recognitions which can be used for dealing with a much wider set of potential problems compared to the originally set, more particular goals.

After this somewhat "mysterious" introduction on the nature of the book, let us tell the story itself. Meanwhile, also the roles played by the members of the contributing team of authors within the research will be clarified.

So, let us see the "story". Several years ago, I was approached by one of our that time Ph.D. students, Adrienn Buruzs, an environmental engineer and lecturer of the corresponding department of the Széchenyi István University in Győr, Hungary. She had a real-life problem to solve, and her intention was to use the solution as an essential part of her future thesis. Her problem was the modeling and analysis of the solid waste management system, originally restricted to the broader region of the city of Győr, a major industrial center in Hungary. The main question was whether the existing management system would be sustainalble, or would collapse in this way, or, maybe, would start chaotic behavior at one point if continued as it had been done for decades.

How do you check the sustainability of a complex procedure where several components affect each other mutually in the most complex way—forming essentially a complete directed graph whose behavior would give an insight into the mutual effects and long-time behavior of these components? Several more traditional approaches of environmental management system analysis attempted prior to the start of our collaboration did not lead to any usable result. Adrienn tried to broaden the scope of her search for methodologies that may give an answer to the basic problem in a satisfactory way, and then she found some papers dealing with FCMs being used in modeling management systems, including the waste management area. Her question was, whether I would accept being her supervisor in applying the FCM model, while she would go on with the environmental issues of the research under the guidance of her former supervisor, and having the help of a team of environmental experts. My answer was positive, and thus our joint research and the story of this book started. This may also explain why in this book we went into so much detail concerning the waste management example—the steps of building up the new models and algorithms would be barely understandable without explaining in sufficient detail all the problems that emerged during the actual modeling approaches.

So, let us begin with the original questions. What is the available input knowledge in this problem? Luckily, for several decades, rather focused attention had been on the waste management issue in Hungary (and in the whole of Europe), and thus, reliable statistical data concerning major aspects of the situation and measures were available. On the other hand, in the frame of a research grant of the Széchenyi István University, a group of environmental experts had been working on the mentioned problem, and knowledgeable estimates of various effects and mutual influences of the components of the management system were also available. So, after taking the decision of attempting the modeling procedure by FCM, we started to look at the models that could be found in the literature; and we soon found there was a rather wide consensus in accepting a certain six-component model, which was considered as the set of factors essentially determining the behavior of all such management systems.

Having the historical data and the model, it was an easy job to build up the first assumedly relevant FCM model and soon it was implemented by one of our colleagues, computer scientist Miklós Hatwagner for doing investigations by simulation. So, we had some results which evidenced that the model would converge, i.e., the waste management system was indeed sustainable in this form. We also had the limit values of the composing factors (concepts, in the FCM terminology), expressed by fuzzy membership degrees in [0, 1], and the bipolar fuzzy graph of the limit FCM also contained the signed membership degrees belonging to the directed edges (in $[-1, 1]$), expressing the sense and intensity of each pairwise influence between the concepts.

At the same time, we also had the estimates of the experts on the directions and absolute values of these degrees of influence. And, to our surprise, those estimates did not match at all with the degrees obtained by running the simulation based on the historical data. There was even no approximate match or similarity. Now, the dilemma raised, whether the experts were totally wrong? Or, were the historical data somehow false? In both cases, we would have to answer with a firm "no". The data came from the very reliable Hungarian Statistical Office and the expert panel was composed of recognized experts and was wide enough to compensate for any individual subjective estimation and thus to prevent making a collective mistake. So what might have been the explanation? At this point, we had to recognize the only way out of this dilemma: the model itself did not fit the reality adequately, as this way understandably, both approaches very likely lead to erroneous results. So, the satisfying results confirming sustainability of the system could not be accepted as reliable and the whole research had to be started from scratch. In this situation, we had to think about constructing our own entirely new model, a model that was more adequate and more sensitive to the complex and intertwined relationships among the influencing concepts. But what concepts? Clearly, the simple six concepts model found in the literature failed. We concluded that it was obviously based on an oversimplification of the problem. So, the next time period was spent collecting more detailed information on the system to model. The panel of experts was anyway available, and the mentioned grant enabled us to organize a special workshop discussing the real concept pool of the problem on hand. The set of potential concepts turned out to be much more numerous than in that simple graph we took from the literature sources. Anyway, the number thus obtained was too high to be overviewed intuitively by any single human expert or even expert team. And, what was even worse, as these concepts were non-standard, no direct statistical data were available from the past.

So, the next task to solve was to create a new technique to extract historical data for the now-introduced new concepts from available documents from and on the period in question. The details of this investigation can be found in the chapter on the expert workshop organized in order to produce input data, a rather successful and—in this context—novel method based on text mining. After this, it was a straightforward work that lead to simulation results concerning the final (convergence) states and the mutual influence directions and degrees of the waste management system.

The next problem, however, emerged then immediately: the new, more detailed model was unambiguously too complex to be handled by the experts involved in this research (or, by any human experts), 33 new concepts which were established by "splitting" the 6 components, meant $33^2 = 1089$ directed edges (assuming that in the FCM model, self-loops were allowed; somewhat less, if assuming the opposite), a number that no one could overview intuitively. For such a complex graph, no expert domain knowledge could be used for verifying the results obtained by the simulation based on the artificially created historical data, varification was possible only by observation and comparison—a very time-consuming procedure. Is there a compromise solution between the two models, the too rough and the too fine model? A graph size where the number of nodes is somewhere between 6 and 33, a graph that still adequately reflects the behavior of the original system but can be overviewed intuitively by a human expert? The answer was obvious: an algorithm had to be developed that could reduce the number of concepts to an acceptably small, but still adequately large number of nodes and directed edges with signed fuzzy membership degrees. At the same time, the essence of the original, finer model had to be preserved.

A starting idea was offered by the classic family of state reduction algorithms that have been in use for many decades by now in Digital Design, especially, for optimizing asynchronous and synchronous sequential circuits by reducing the number of states. These algorithms are based on the properties of symmetric binary relations, namely, of equivalence and tolerance relations where reflexivity, symmetry, and in the case of the former, transitivity are present. These relations help establish disjoint equivalence and potentially overlapping compatibility classes, where each such class could take over the role of all the states merged into the given class, while the connecting edges representing transitions could be entirely merged. Of course that way there is a loss of information, but there is no free lunch! These relations had been extended to fuzzy relations constructing fuzzy classes with truth values between 0 and 1, representing the degrees of belonging to the classes. In this approach, the concepts could be handled as the states in the examples, while the edges with signed degrees of mutual influences were comparable to the state transition edges. The difficulty in this approach lies in finding a proper metrics, where the merger of edges would satisfy the conditions of being "similar", i.e., remaining within certain limits compared to all included effects between pairs of concepts, and the merger of the nodes also underlie similar conditions. You will realize when reading the respective chapter that it was not at all easy to find such appropriate metrics.

Once obtaining some good candidates for the potential reduction algorithms for the concepts, the next problem was how to prove, or at least how to find strong evidence that after the reduction, the thus merged concepts would still have a meaning in the context of the original problem, and that the new degrees of mutual influences would be sufficiently close to the expected merged influences. As an exact mathematical proof seemed to be hopeless, we started very extensive simulational experiments, both on real-life sytems found in the literature, and on a large number of randomly generated artificial examples. In this very tiresome work that was led by Miklós, he received help from a team of international researchers whom he could collaborate with during a study leave financed by the Széchenyi University. The two names to mention here in the first place are Elpiniki Papageorgiou and Engin Yeşil.

The results were promising, so we started to apply the new algorithm on the original problem, obtaining various size reductions of the extended concept model of the waste management system, and so we obtained results rather in accordance with expert knowledge. A somewhat surprising side effect of applying the reduction algorithm to our engineering case study was that after splitting the original six concepts, and then merging them again into tolerance classes, i.e., reduced concepts, the mergers did not follow the starting concepts at all, but often "split concepts" belonging to different "mother concepts" were merging together, which fact clearly showed that the original simple model was essentially wrong. Some split concepts played surprisingly important role in forming several, even, many of the reduced classes, a fact that stresses the special significance of these "component concepts" in the behavior of the whole system.

This research had some challenges, especially the procedure of searching for proper metrics (for the given model, and the given problem) offered some surprises, even some disappointments concerning the initial expectations. The results of these investigations, with some conclusions and much supporting evidence for the eventually chosen distance measures' applicability, can be found in the respective chapters of the book.

And the story went on... As a matter of course, we had published the results at various conferences and in some journal papers, and that brought the attention of some peers to our group's activities in modeling real-life management systems by FCM. At one point, my old friend Vesa Niskanen contacted me and proposed that we collaborate in a project run by the French École Central for a bank where the behavior of a set of management models for various bank-related systems had to be taken under investigation. The problem had some connections with the subject of our previous research, however, here no historical data, only estimates of the senses and degrees of mutual influences among the concepts were available. Instead of historical data, the present states were also assessed by those experts, so the initial conditions were given.

The weak points of the model could be identified first of all in the uncertainty of the estimates, both concerning the mutual effects and the initial states. As the goal of the modeling procedure was to establish stable operational states in the bank systems under investigation, the key question was here similar to the waste management-related case: whether the bank management system would converge to a fix point attractor under given conditions, i.e., whether it would stabilize itself in a sustainable operation, or, it would get into a periodical, instable, or even chaotic behavior. What is the key problem here? While the experts' estimates are obviously subjective and contain vague components, the key issue is if the model is adequate and if its sensitivity for the uncertain elements does not cause any non-deterministic behavior, namely, instable limit behavior that is dependent on the initial states, the parameters of the model, and the uncertainty degrees of the expert estimates.

This project was the start of a new direction of our investigations, the global analysis of the stability and limit behavior of FCM models. The research went on in two streams. Miklós F. Hatwágner was the leader of extended simulational experiments, first studying the bank model given in the project, later opening the view to a rather wide scope of possible models. István Á. Harmati, the mathematician team member joined our group and started to carry out stability and convergence analysis of the models. The results obtained clearly showed that FCMs are sensitive tools, and parameters of the system (reflecting the speed and interaction chain of the state changes, and the accuracy of the estimations of states and mutual effect values, largely influence the behavior of the system, even, the number of fix point attractors, or the presence of chaotic behavior may depend in a rather sensitive way on the mentioned aspects. We trust the results thus obtained will help other researchers design and analyze similar models for a wide variety of applications.

As it goes, the mathematical analysis brought some further ideas. As the uncertainty of the elements in the model and of the values clearly plays a crucial role in the stability, we started to focus on various uncertain cognitive map models, including other types beyond simply fuzzy CM-s. The last part of the book will show the results in this respect, especially investigating gray fuzzy extensions. Mathematical results concerning the limit behavior of such generalized FCMs, including sufficient conditions of the existence of (a single) fix point attractor will be found in the last chapters.

It is obvious that the story has to go on further. All the sub-topics discussed in this book raise new questions and only some of the answers to these questions are given in this volume. Our group is continuing the work and we encourage fellow FCM researchers to join us or start connected investigations—the floor is open.

At the end of the Preface, I would like to mention that all results in the book are due to collaborative research, but in every case, there emerges a "chief investigator", the leader of that particular research task. In every chapter, only one name, the name of the main contributor of the given part of the work, the responsible "chief investigator" was indicated. My name as the name of the leader of the whole team does not appear in any of the chapters as my name is anyway, on the book cover, and I advised and led the collaborators to some degree in all the sub-tasks and during the whole project.

At the end of the Preface, I want to repeat my hope that the book will be of some interest to the reader and that it will provide useful knowledge, both in the technical sense, namely, how to construct models by fuzzy cognitive maps and extended FCMs, and in the methodological sense, presenting, how a concrete problem may lead to a chain of questions with scientific aspects, how answering some of these questions may bring further interesting questions and recognitions, and how the available (engineering or management) expert domain knowledge may help the model builder and user, the computer scientist and the mathematician, in solving real-life problems efficiently, while adding to the pool of abstract knowledge at the same time.

Enjoy the book!

Győr, Hungary
2020

László T. Kóczy

Acknowledgements This work was funded by the Hungarian Office for Research Innovation and Development (NKFIH), grant number K124055.

Contents

Chapter 1
Introduction to Fuzzy Cognitive Maps

Miklós F. Hatwagner

1.1 The Birth of Fuzzy Cognitive Maps

Cognitive Maps (CM) are used in political analysis and decision making in international relations, foreign policy for a long time. The method was suggested by Robert Axelrod in his book [1] in the late '70s. According to Bart Kosko's description in [17], these maps are signed digraphs. Graphs, as algebraic structures have two components: nodes and edges (arcs). In CM, nodes represent variable *concepts* (eg. social instability) and the causal connections among these concepts are characterized by edges. The edges have a direction and a sign. If concept A causally increases concept B, it is represented by an edge from A to B with positive sign. On the other hand, if A reduces the value of B, the edge has a negative sign. Kosko illustrated CM with an example based on Henry Kissinger's essay "Starting Out in the Direction of Middle East Peace" published in *Los Angeles Times* in 1982 (see Fig. 1.1). Besides the graph, he also composed the adjacency (connection, weight) matrix (Fig. 1.2) of the model. Only three different values can be found in this matrix, representing the causal relationship among concepts. If $w_{ij} = w(C_i, C_j)$ is 1, concept C_1 causally increases the value of C_2 (positive edges). On the contrary, if C_1 causally decreases the value of C_2, it is represented by -1 (negative edges), and the value 0 indicates the lack of causal connection.

It became quickly evident that the structure of a Cognitive Map imply too much limitations. The degree of causality, the levels of the causal effects (sometimes–often, little–much, etc.) cannot be expressed with the existing tool, and needs further development. Kosko introduced the Fuzzy Cognitive Map (FCM) [17], where the edges may have several causality values. This way the original CM turned into a bipolar fuzzy graph [31]. Kosko also developed a fuzzy causal algebra for propagating causality in it, making the static analysis of the model possible. The concepts may affect other concepts indirectly because of the cyclic and non-feedforward structure of FCM.

L. T. Kóczy, *Fuzzy Cognitive Maps*, Studies in Fuzziness and Soft Computing 427,
https://doi.org/10.1007/978-3-031-37959-8_1

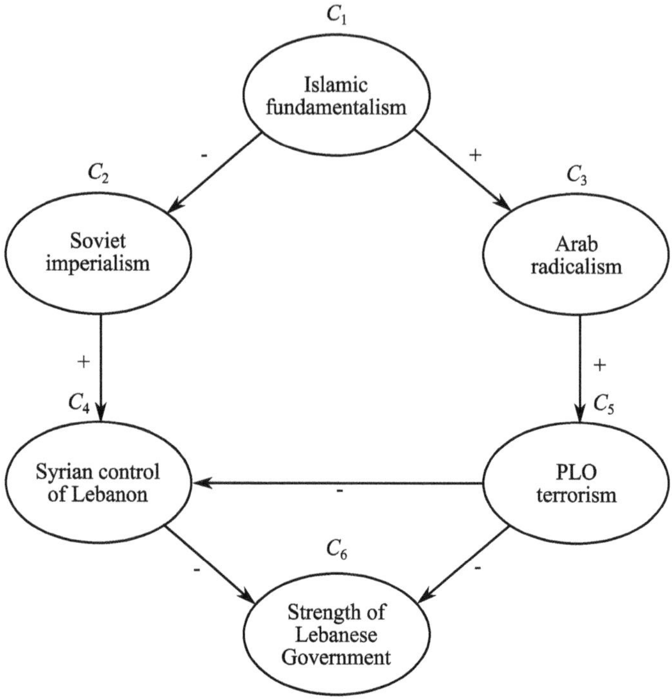

Fig. 1.1 The Cognitive Map drawn by Kosko based on Kissinger's essay

Fig. 1.2 The adjacency matrix of the CM based on Kissinger's essay

$$\begin{array}{c} \\ C_1 \\ C_2 \\ C_3 \\ C_4 \\ C_5 \\ C_6 \end{array} \begin{array}{c} \begin{array}{cccccc} C_1 & C_2 & C_3 & C_4 & C_5 & C_6 \end{array} \\ \begin{pmatrix} 0 & -1 & 1 & 0 & 0 & 0 \\ 0 & 0 & 0 & 1 & 0 & 0 \\ 0 & 0 & 0 & 0 & 1 & 0 \\ 0 & 0 & 0 & 0 & 0 & -1 \\ 0 & 0 & 0 & -1 & 0 & -1 \\ 0 & 0 & 0 & 0 & 0 & 0 \end{pmatrix} \end{array}$$

1.2 Simulations

The visual representation and formal description of models may help experts to review the structure of the studied system, but the real benefit of FCMs is the possibility of running simulations. This can be done by dynamic analysis [18]. Using a simple inference technique, the next state of concepts (also called the *activation values* of concepts, based on the similarity of FCMs and artificial neural networks) can be calculated using their current state and the weight of connections among them [4, 28]. This way what-if analysis can be performed which is very useful for decision makers.

1.2.1 Inference Rules

Kosko defined the first, and still widely applied inference rule (Eq. 1.1). In this case concepts are not allowed to directly influence themselves. With other words, a concept does not have "memory", technically speaking the edges must connect different concepts (i.e. self-loops are forbidden), but indirect feedbacks are still possible. Accordingly, the main diagonal of the connection matrix contains only zeros. These models are called FCMs of Type I [27].

$$A_i^{t+1} = f\left(\sum_{\substack{j=1 \\ j \neq i}}^{n} w_{ji} A_j^t\right) \tag{1.1}$$

Here, A_i^{t+1} is the activation value of concept i at time step $t+1$, n is the number of concepts and $f(\cdot)$ is the *threshold* (transformation, squeezing) function. Without this function, the activation values of concepts may exceed or fall below their allowed extreme values.

FCMs of Type III allow concepts to take into account their current states during the calculation of the next state, and apply the inference rule defined by Eq. 1.2. In this case, the main diagonal of the connection matrix may contain nonzero values, which are often ones in practice.

$$A_i^{t+1} = f\left(w_{ii} A_i^t + \sum_{\substack{j=1 \\ j \neq i}}^{n} w_{ji} A_j^t\right) \tag{1.2}$$

FCMs of Type II also have a one step memory, but the weight of self-loops (k_1) are the same for all concepts. Moreover, the influence coming from other concepts can be limited as well (k_2, see Eqs. 1.3, 1.4).

$$A_i^{t+1} = f\left(k_1 A_i^t + k_2 \sum_{\substack{j=1 \\ j \neq i}}^{n} w_{ji} A_j^t\right) \tag{1.3}$$

$$0 < k_1, k_2 \leq 1 \tag{1.4}$$

Papageorgiu suggested the re-scale inference rule (Eq. 1.5) in [23], because problems emerged in cases where the initial values of concepts were 0 or 0.5, assuming that the initial states are known at all. This modified inference rule yielded more reliable and natural results than other FCM versions.

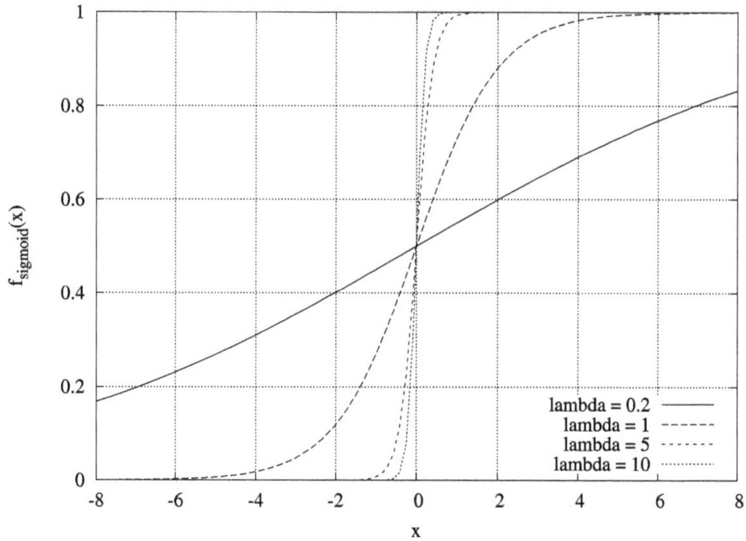

Fig. 1.3 Steepness of the sigmoid function with various λ parameters

$$A_i^{t+1} = f\left((2A_i^t - 1) + \sum_{\substack{j=1 \\ j \neq i}}^{n} w_{ji} \cdot (2A_j^t - 1)\right) \tag{1.5}$$

1.2.2 *Threshold Functions*

In general, the activation values are in the $A_i \in [0, 1]$ interval. Several threshold functions were published during the years, but the most widely applied one is the *sigmoid* (*logistics*) function (Eq. 1.6):

$$f_{\text{sigmoid}}(x) = \frac{1}{1 + e^{-\lambda x}} \tag{1.6}$$

where the $\lambda > 0$ specifies the steepness of the function. It's typical value is 5. With greater values it approximates a binary function, with lower values a linear function (see Fig. 1.3).

Choosing an appropriate λ value

Is there an "appropriate" value of this parameter? According to Bueno and Salmeron [3], the appropriate value should be found during an experimentation phase. As it was

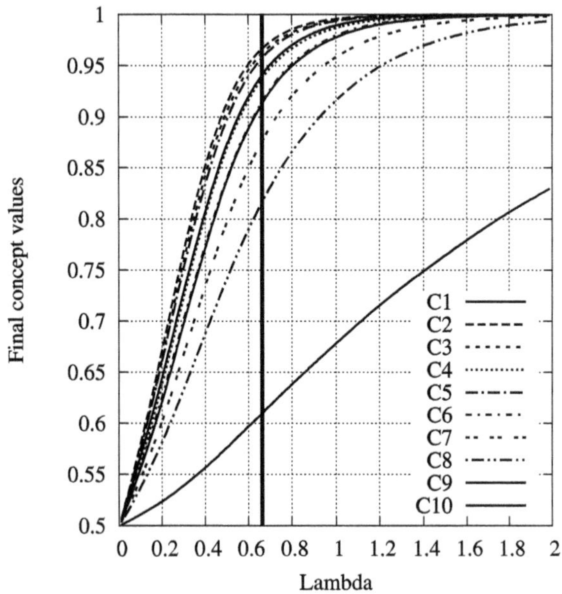

Fig. 1.4 The final concept values of a Stakeholder Relationship Management System as function of parameter λ. $\lambda = 0.664$ maximized the standard deviation of concept values (marked with the thick vertical line)

pointed out in [9, 13], the congergence speed and the final, stable values of concepts are heavily affected by the value of parameter λ (see Fig. 1.4). Despite of the different results, different λ values does not change the result of a simulation qualitatively, because the sigmoid function is strictly monotonically increasing. In practice, it is still worth to find an "optimal" value for λ, because sometimes the otherwise hardly different concept states may become distinguishable, and the order of concepts can be defined, if it is required. A possible method for "lambda optimization" is to maximize the standard deviation of final concept values (Fig. 1.5) even by a quick local search algorithm, eg. the Golden Section Search.

Even if the value of λ can be considered "optimal", a scalar value is unable to unleash the full modeling potencial of an FCM. In an extreme case, every concept may have its own, specific λ value, in order to get more realistic predictions. This way, the inference rule defined by Eq. 1.1 can be re-written to Eq. 1.7, where f_i represents the threshold function of concept C_i that uses λ_i. This approach increases the number of model parameters, however. One of the benefits of using FCM was its simple structure, which can be understood easily by stakeholders, experts of different fields, and the values (initial states of concepts, connection weights) can be matched to real values or effects in a straightforward way. The values of λ do not have trivial connection with any real world objects, however, and their values can be set with trial and error technique or computationally. It also increases the required computational power of applications because it increases the number of model parameters. A good balance can be found if some concepts have their own parameters, the others share a common value.

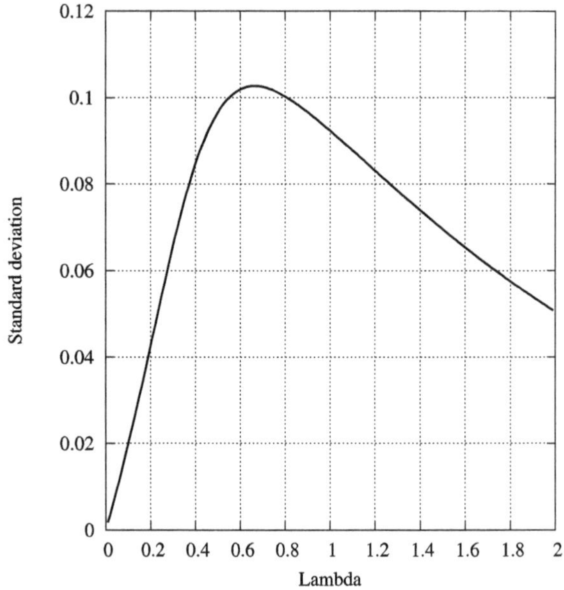

Fig. 1.5 Standard deviation of final concept values of a Stakeholder Relationship Management System calculated with different λ values

$$A_i^{t+1} = f_i\left(\sum_{j=1}^{n} w_{ji} A_j^t\right) \tag{1.7}$$

Concepts may be classified in three groups: if a concept influences the state of other concepts, but not affected by other concepts, it can be called an *input* concept. Depending on the applied inference method, these concepts keep their initial values or their values can be sustained artificially (like "sustained inputs" in [4]), therefore they do not need threshold function and parameter λ neither. Another group of concepts is affected by other concepts but its elements do not influence the state of other concepts. In many applications, only these concepts are used for decision making, and they are called *output* concepts. The remaining concepts are both affected by and they also affect other concepts, and are called *intermediate* concepts [2].

If all output concepts have specific λ values, and intermediate concepts have a common λ, the complexity of the model remains moderate but its forecasting ability can be increased as it happened in [12]. If historical time series data are available, λs can be optimized with eg. the Big Bang—Big Crunch algorithm [29].

The *sign* function (Eq. 1.8) is also a popular threshold function, but because of the two available values, a concept can only be activated (1) or deactivated (0).

$$f_{\text{sign}}(x) = \begin{cases} 1, & x > 0, \\ 0, & x \leq 0 \end{cases} \tag{1.8}$$

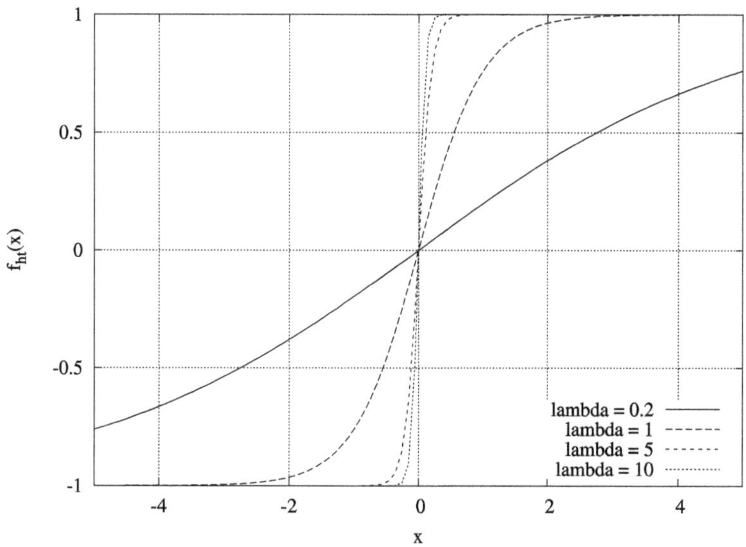

Fig. 1.6 Steepness of the hyperbolic tangent function with various λ parameters

In some cases the system to be modeled requires the activation values to be in $A_i \in [-1, +1]$, and different threshold functions have to be applied. If continuous states are allowed, the *hyperbolic tangent* function (Eq. 1.9) is a possible choice:

$$f_{\mathrm{ht}}(x) = tanh(\lambda x) = \frac{e^{\lambda x} - e^{-\lambda x}}{e^{\lambda x} + e^{-\lambda x}} \tag{1.9}$$

The λ parameter can be used here as well, and has similar effect on the steepness of the function than on sigmoid function's (Fig. 1.6).

If only discrete states have to used, the *trivalent* function (Eq. 1.10) can be the solution. The value 1 expresses the increasing, the value -1 expresses the decreasing and 0 means the stable state of a concept. A slightly modified version of this function can be found in [30] (Eq. 1.11).

$$f_{\mathrm{trivalent1}}(x) = \begin{cases} 1, & x > 0, \\ 0, & x = 0, \\ -1, & x < 0. \end{cases} \tag{1.10}$$

$$f_{\mathrm{trivalent2}}(x) = \begin{cases} 1, & x \geq 0.5, \\ 0, & -0.5 < x < 0.5, \\ -1, & x \leq -0.5. \end{cases} \tag{1.11}$$

1.2.3 Simulation Outcomes

During simulations, the states of concepts continuously evolve, until one of the three possible outcomes happens [28]:

1. In most cases, the activation values converge very quickly to a stable state (equilibrium point, see Eq. 1.12). These states are also called *fixed points* (FPs). Depending on the connection matrix, initial state, inference rule, threshold function and its λ parameter (if applied), a model may have multiple FPs.

$$\exists t_{\mathbf{FP}} \in \mathbb{N} : A_i^{t+1} = A_i^t, \forall t \geq t_{\mathbf{FP}}, i = 1, \ldots, n \tag{1.12}$$

2. Sometimes the system enters a *limit cycle* (LC). In these cases after several time steps, a T steps long sequence of state vectors starts to repeat (see Eq. 1.13).

$$\exists t_{\mathbf{LC}}, T \in \mathbb{N} : A_i^{t+T} = A_i^t, \forall t \geq t_{\mathbf{LC}}, i = 1, \ldots, n \tag{1.13}$$

3. The system never stabilizes, behaves chaotically.

It's easy to prove that bivalent and trivalent systems cannot behave chaotically. The process of simulations are deterministic, and the number of different state vectors is 2^n or 3^n, respectively, where n is the number of concepts. According to this, if the system is unstable, its states are going to repeat after 2^n or 3^n time steps, or earlier. FPs and LCs are related to the hidden patterns described in Kosko's work [18].

In contrast, continuous state FCMs may enter practically infinite different states, thus chaotic behavior is possible. It happens especially often in case of Type III FCMs, which enter LCs frequently as well [21]. Martchenko et al. also defined the condition of stability, too, which depends on the connection matrix.

Harmati et al. went further [6] and determined the conditions of convergence and the number of FPs. According to their results, if the FCM uses sigmoid threshold function and the inequality

$$||W||_F < \frac{4}{\lambda} \tag{1.14}$$

holds, then the FCM has exactly one FP, without any LCs or chaotic patterns. Here W is the connection matrix, $||\cdot||_F$ is the Frobenius-norm of the matrix defined by Eq. 1.15, and λ is the steepness of the threshold function.

$$||W||_F = \left(\sum_i \sum_j w_{ij}^2 \right)^{1/2} \tag{1.15}$$

Similarly, if the hyperbolic tangent function is applied and Eq. 1.16 holds, the FCM has one and only one FP. These theorems are valid even if the connection matrix contains direct feedbacks.

$$||W||_F < \frac{1}{\lambda} \tag{1.16}$$

Moreover, if the sigmoid or hyperbolic tangent threshold functions are used, and W contains only non-negative elements than the FCM has FP, regardless of the value of λ. Similar contexts were also explored in case of some extended versions of FCM, the Grey FCMs [7] and Fuzzy Set Valued Sigmoid FCMs [8].

These are remarkable achievements because some of the behavioral properties of an FCM can be revealed without executing an exhaustive set of simulations, which is computationally very intensive. Unfortunately it is often unavoidable. If the model is made by experts, it may contain many subjective elements, eg. it is really hard to define the exact weight of a relationship. Even a small difference in the weights may lead to different behaviour, however. It is worth investigate the effect of small model modifications on behavioural properties, but it means a huge job which must be automated like in [10, 11]. The search space is huge for an exhaustive search hence Bacterial Evolutionary Algorithm [22] was applied to generate and analyse the behavioral difference caused by slightly modified model versions and λ values. If modified models with significantly different dynamic behaviour are found, experts are suggested to consider some changes on the model.

In most decision making tasks, stable states are the desired results. In these cases the future state of a system can be determined, and if that equilibrium state is undesirable, measures can be taken. There are some situations however, eg. time series predictions [14, 15, 20], when LCs or chaotic behavior seems very useful. Often extended versions of the original FCM are suggested to get better predictions.

1.2.4 An Example Simulation

Finally, an example FCM is given for illustration. The model has 5 concepts, their allowed values are in the interval [0, 1]. Kosko's original inference rule is applied (Eq. 1.1), the threshold function is the sigmoid type (Eq. 1.6) with parameter $\lambda = 5$.

Formally, this model can be defined by a 4-tuple (C, W, A, f), where $C = C_1, C_2, \ldots, C_n$ is the set of $n = 5$ concepts, function $W : (C_i,\ C_j) \to w_{ij}$ associates the causal value $w_{ij} \in [-1,\ +1]$ to each pair of concepts $(C_i,\ C_j)$. These weights of directed edges are collected in the connection matrix denoted by W_{nxn} (see Table 1.1 and Fig. 1.7). Another function, $A : (C_i) \to A_i$ associates the activation value $A_i \in \mathbb{R}$ to each node C_i at each time step $t = 1, 2, \ldots, m$. The (sigmoid) threshold function $f : \mathbb{R} \to [0,\ 1]$ forces the activation values in their allowed range.

These are very general settings, and according to this, the dynamic behavior of the model is also very typical: the concepts' values converge after some time steps to their final values, to an equilibrium point (see Table 1.2 and Fig. 1.8).

Table 1.1 Connection matrix of the example FCM

	C_1	C_2	C_3	C_4	C_5
C_1	0	0.5	0.3	0	0
C_2	0	0	0	−0.1	0.35
C_3	0	−0.1	0	−0.2	0.4
C_4	0.4	0	0	0	0
C_5	−0.7	0	0	0	0

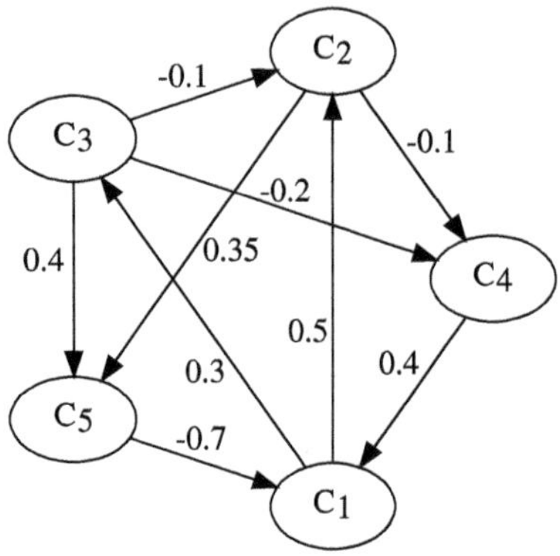

Fig. 1.7 Graphical representation of the example FCM

Table 1.2 State vectors of the example FCM

Time	C_1	C_2	C_3	C_4	C_5
t_0	0.00	0.50	1.00	0.50	0.00
t_1	0.94	0.44	0.32	0.50	0.50
t_2	0.83	0.65	0.57	0.87	0.04
t_3	0.92	0.41	0.25	0.84	0.05
t_4	0.80	0.42	0.28	0.86	0.04
t_5	0.81	0.41	0.27	0.83	0.06
t_6	0.81	0.42	0.28	0.84	0.05
t_7	0.81	0.42	0.28	0.83	0.06
t_8	0.81	0.42	0.28	0.84	0.05
t_9	0.81	0.42	0.28	0.84	0.05

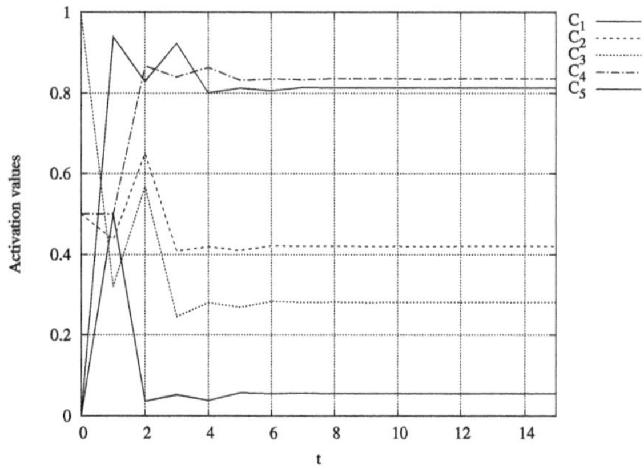

Fig. 1.8 Convergence of concept activations during simulation

1.3 Creating a Fuzzy Cognitive Map Model

There are two main ways of model creation: the first one is called *deductive modeling*, when experts are asked to identify the main driving factors (concepts) and the cause-effect relationships among them (connection matrix). These techniques do not require the existence of recorded historical data, because they utilize the domain-specific knowledge of experts. Unfortunately, these experiences are often limited, contain subjective opinions as well, and only in case of relatively small systems can be modeled this way with high confidence.

The other way is called *inductive modeling*, when objective, measured historical data are used to generate the model with learning algorithms. These algorithms do not require human interaction in general, but the concepts are still often defined by experts.

It is often hard to define which concepts play an important role in the model. According to the strategy described in [26] and originally suggested in [16, 17] all available concepts should be included in a list, and later remove the insignificant ones. A similar approach is suggested, but executed automatically in [12]. The hardest job for experts is the definition of connection weights among concepts, however, because the number of connections is the quadratic function of the number of concepts.

In many cases the capability of exact re-generation of the time series used for learning is not a requirement. If the model assigns initial states and their corresponding fixed points correctly, that is enough in most decision making tasks.

Some possible methods of model creation are suggested by Kosko in his early papers (eg. in [4, 18]). In order to increase the credibility of experts, he advised to ask several experts, then combine and differentiate their suggestions based on their credibility. Groumpos described this technique and provided an alternative one in [5]. Kosko also suggested a simple learning technique based on the Hebbian law (DHL). Later several authors improved his method (eg. NHL, AHL, DD-NHL). A good summary of these are available in [24], which presents some population-based methods as well. The book [25] gives a wider insight into model creation methods, including hybrid techniques. Even today are new methods born, eg. [19] takes a new approach.

Acknowledgements The research presented in this paper was carried out as part of the EFOP-3.6.2-16-2017-00016 project in the framework of the New Széchenyi Plan. The completion of this project is funded by the European Union and co-financed by the European Social Fund.

References

1. Robert A (1976) Structure of decision: the cognitive maps of political elites. Princeton University Press, Princeton, NJ
2. Yiannis B, Christodoulou MA, Theodoridis D, Kottas T (2014) System identification and adaptive control. Theory and applications of the neurofuzzy and fuzzy cognitive network models
3. Bueno S, Salmeron JL (2009) Benchmarking main activation functions in fuzzy cognitive maps. Expert Syst Appl 36(3):5221–5229
4. Dickerson JA, Kosko B (1994) Virtual worlds as fuzzy cognitive maps. Presence: Teleoperators Virtual Environ 3(2):173–189
5. Groumpos PP (2010) Fuzzy cognitive maps: basic theories and their application to complex systems. In: Fuzzy cognitive maps, pp 1–22. Springer
6. Harmati IÁ, Hatwágner MF, Kóczy LT (2018) On the existence and uniqueness of fixed points of fuzzy cognitive maps. In: International conference on information processing and management of uncertainty in knowledge-based systems, pp 490–500. Springer
7. Harmati IÁ, Kóczy LT (2018) On the convergence of fuzzy grey cognitive maps. In: Conference on information technology, systems research and computational physics, pp 74–84. Springer
8. Harmati IÁ, Kóczy LT (2018) On the existence and uniqueness of fixed points of fuzzy set valued sigmoid fuzzy cognitive maps. In: 2018 IEEE international conference on fuzzy systems (FUZZ-IEEE), pp 1–7. IEEE
9. Hatwagner MF, Koczy LT (2015) Parameterization and concept optimization of FCM models. In: 2015 IEEE international conference on fuzzy systems (FUZZ-IEEE), Istanbul, Aug 2015. IEEE, pp 1–8
10. Hatwagner MF, Vastag G, Niskanen VA, Kóczy LT (2018) Improved behavioral analysis of fuzzy cognitive map models. In: International conference on artificial intelligence and soft computing pp 630–641. Springer
11. Hatwágner MF, Vastag G, Niskanen VA, Kóczy LT (2019) Banking applications of FCM models. In: Trends in mathematics and computational intelligence. Springer, pp 61–72
12. Hatwágner MF, Yesil E, Dodurka MF, Papageorgiou, E., Urbas L, Kóczy LT (2018) Two-stage learning based fuzzy cognitive maps reduction approach. IEEE Trans Fuzzy Syst 26(5):2938–2952
13. Hatwágner MF, Torma A, Kóczy LT (2015) Parameter dependence of fuzzy cognitive maps' behaviour. In: 2015 10th Asian control conference (ASCC), May 2015, pp 1–6
14. Homenda W, Jastrzebska A (2017) Clustering techniques for fuzzy cognitive map design for time series modeling. Neurocomputing 232:3–15
15. Homenda W, Jastrzebska A, Pedrycz W (2014) Modeling time series with fuzzy cognitive maps. In: 2014 IEEE international conference on fuzzy systems (FUZZ-IEEE). IEEE, pp 2055–2062
16. Khan MS, Quaddus M (2004) Group decision support using fuzzy cognitive maps for causal reasoning. Group Decis Negot 13(5):463–480
17. Kosko B (1986) Fuzzy cognitive maps. Int J Man-Mach Stud 24:65–75
18. Kosko B (1988) Hidden patterns in combined and adaptive knowledge networks. Int J Approx Reason 2(4):377–393

19. Lu W, Feng G, Liu X, Pedrycz W, Zhang L, Yang J (2019) Fast and effective learning for fuzzy cognitive maps: a method based on solving constrained convex optimization problems. IEEE Trans Fuzzy Syst
20. Wei L, Yang J, Liu X, Pedrycz W (2014) The modeling and prediction of time series based on synergy of high-order fuzzy cognitive map and fuzzy c-means clustering. Knowl-Based Syst 70:242–255
21. Martchenko AS, Ermolov IL, Groumpos PP, Poduraev JV, Stylios CD (2003) Investigating stability analysis issues for fuzzy cognitive maps. In: 11th mediterranean conference on control and automation-MED'03
22. Nawa NE, Furuhashi T (1998) Bacterial evolutionary algorithm for fuzzy system design. In: SMC'98 conference proceedings. 1998 IEEE international conference on systems, man, and cybernetics (Cat No 98CH36218), vol 3. IEEE, pp 2424–2429
23. Papageorgiou EI (2011) A new methodology for decisions in medical informatics using fuzzy cognitive maps based on fuzzy rule-extraction techniques. Appl Soft Comput 11:500–513
24. Papageorgiou EI (2012) Learning algorithms for fuzzy cognitive maps—a review study. IEEE Trans Syst Man Cybern Part C (Applications and Reviews) 42(2):150–163
25. Papageorgiou EI (2013) Fuzzy cognitive maps for applied sciences and engineering: from fundamentals to extensions and learning algorithms, vol 54. Springer Science & Business Media
26. Stach W, Kurgan L, Pedrycz W (2010) Expert-based and computational methods for developing fuzzy cognitive maps. In: Glykas M (ed), Fuzzy cognitive maps—advances in theory, methodologies, tools and applications. Springer, pp 23–42
27. Stylios CD, Groumpos PP et al (1999) Mathematical formulation of fuzzy cognitive maps. In: Proceedings of the 7th mediterranean conference on control and automation, pp. 2251–2261
28. Tsadiras AK (2008) Comparing the inference capabilities of binary, trivalent and sigmoid fuzzy cognitive maps. Inf Sci 178(20):3880–3894 (2008)
29. Yesil E, Urbas L (2010) Big bang-big crunch learning method for fuzzy cognitive maps. World Acad Sci Eng Technol 71:816–825
30. Yesil E, Urbas L, Demirsoy A (2014) FCM-GUI: a graphical user interface for big bang-big crunch learning of FCM. In: Fuzzy cognitive maps for applied sciences and engineering. Springer, pp 177–198
31. Zhang W-R (1998) Bipolar fuzzy sets. In: Fuzzy systems proceedings, 1998. The 1998 IEEE international conference on IEEE world congress on computational intelligence, vol 1. IEEE, pp 835–840

Chapter 2
Systemic Relationships Between Regional Sustainability and Waste Management

Adrienn Buruzs

2.1 Introduction

In the next, four chapters on the case study of the original environmental engineering problem, namely, modeling waste management systems will be presented, in order to explain in detail what was the motivation for introducing a family of new models and algorithmic approaches in the FCM modeling area. Even though these chapters will present a special application problem and the phases of getting closer to its solution, they may be interesting for the Reader who focuses on FCMs only, as a real life case study helps realize how step by step a problem may be solved with a Computational Intelligence tool. Let us start thus with the engineering problem.

First, we will look at the reasons and circumstances that result in the generation of waste. Consumption is an integral part of life. We consume the oxygen of the air, the water of rivers, the produce of nature, the animals and the treasures of the earth. As we consume, we transform them into forms that are finer, more useful. Eventually, these goods return to nature, generating some waste, which again attempts to transform them into energy levels corresponding to their own logic. The cycle has been going on for millions of years, but in the last few centuries man has been testing the circularity of the Earth and nature, giving more and more responsibilities to the higher regulation of balance.

Researchers of later ages have been greatly helped by the analysis of the "waste" left by our human ancestors in understanding the history of mankind. A given society and its development are well characterized by its waste. Since the existence of manufacturing and industry, industrial waste has always been produced, but from an environmental point of view, in modern times, it was only considered a moderate problem as it is locally concentrated, it is of known composition, so it may be only a matter of money, technology and proper legislation to dispose of it or to recycle it as secondary material.

Municipal waste, being of less amount but mixed in the components is, however, much more problematic. Especially, as the amount of municipal waste is increasing worldwide due to rapid population growth, changing consumption patterns and

L. T. Kóczy, *Fuzzy Cognitive Maps*, Studies in Fuzziness and Soft Computing 427,
https://doi.org/10.1007/978-3-031-37959-8_2

associated industrial production. Improperly treated solid waste poses a risk to human health and the environment. Uncontrolled waste disposal and improper waste disposal can also cause damage to environmental compartments, wildlife and human health. Therefore, careful planning and implementation of waste management requires extensive programming. Well-designed and developed waste collection, transport, and disposal programs can (partly) eliminate these problems.

Environmental protection, including waste management, has a relatively short economic, social and legal sphere, and consequently, methods of independent analysis, evaluation and planning have not yet developed [4]. In order to make up for this deficiency as easily as possible, it is necessary to examine the applicability—subject to the specific features—of disciplines [11] that have proved useful in identifying and balancing the sustainability characteristics of other similar management issues.

An assessment of local needs and conditions is essential to support an efficient waste management system, and the waste management methods and components that best suit these conditions can be selected. Each of these activities requires careful planning in terms of financing, waste collection and transportation, selection of processing and disposal technology, regulatory requirements and the needs and capabilities of the system users (the general public). Sustainable municipal waste management offers a comprehensive interdisciplinary framework for waste management. A key factor in planning is the involvement of all stakeholders (citizens, public service providers, municipalities, authorities, research centers, NGOs, etc.) in decision-making.

Creating a sustainable model for municipal waste services is a challenge for local governments.

In addition to the review of methods for the comprehensive analysis of regional development, including waste management, the aim of the research in the case study was to develop and present a methodology that could be applied in practice to assess the sustainability of regional municipal waste management systems, whether from the .mathematical point of view stable behavior of the whole waste management system could be achieved.

Numerous studies have addressed the feasibility of treating municipal waste, and the interactions between key elements of waste management systems in general, but researchers have paid relatively little attention to dis-aggregating factors and to the examination of their interactions with each other. The primary aim of the research group was to identify the conditions under which the municipal solid waste management system could be developed and operated effectively. Understanding the waste management procedures is the key to understanding the impact that different policies and legislation, funding options, population motivation, and available technologies may and will have on waste management performance.

The management of municipal solid waste is shaped by the legislation containing respective regulations at the EU, national and municipal levels. Although in most cases the legislation has contributed to a high level of municipal solid waste management, some of these laws have undoubtedly caused significant disruption to the systems [13]. The key to achieving a sustainable society is choosing the best way to manage waste in every way.

2.2 The Methodological Approach—A Brief Overview

As a first step in the research, we have reviewed the relevant domestic and international literature in order to establish the theoretical foundations. The "knowledge base" thus acquired was complemented by additional information obtained from informal discussions at domestic and international conferences in the field.

After an overview of various alternative approaches of modeling the problem, it was decided that the Fuzzy Cognitive Map (FCM) method seems to be the most suitable, and thus it was chosen for the analysis of the stability (sustainability).

In the first phase of the research, we started from the assumption that the six main factors identified and generally accepted in the literature for modeling municipal waste management are adequate. Indeed, there is strong consensus in the relevant literature that the typical regional waste management system (RWMS) contains (at least) these six major factors. According to that consensus these factors are the "key drivers" of sustainable RWMS, which sufficiently describe and essentially determine why the system under investigation operates in the observed way.

An input data set that served as starting point was based on observations, which could be considered a relatively objective technique. The trend of the examined factors was based on the analysis of literature data. A retrospective reconstruction of their weights and roles was then performed by analyzing the historical data of the factors.

In order to gain expert knowledge, the group chose the method of online questionnaire survey. The purpose of the questionnaire was to identify the impact of each of the determinants of waste management systems and the extent of the impact, based on feedback from professionals. This allows you to easily illustrate the strength of the interaction of factors. Subsequently, fuzzy graph structure model calculations were made to produce a number of variants, which allowed the evaluation of each factor to be independent of subjective judgment.

Despite the fact that waste management systems are supposed to consist of the six main factors identified in the previous approaches, by the end of the first phase of the study it became clear that some novel system-based model was needed in order to review the whole process. Such a new model should be suitable for investigating factors (components) of fundamentally different types of system components, where collaboration and invisible relationships are also required.

In the second phase of the research, an expert workshop was organized on the basis of the system-of-systems approach [2]. The aim of the workshop as an analytical method was to develop a more complex model allowing as many factors (concepts) as necessary in order to properly describe the waste management system. The method applied during this workshop was to first "split" the six main concepts and this way to obtain a considerably larger number of factors. The result proposed by the body of experts participating was a 33-factor model of the Regional Waste Management System (RWMS). At this stage, we became aware of how experts perceive and evaluate future RWMS perspectives and risks in the light of factors.

However, producing time series of the thus identified 33 factors, whether based on literature or expert judgment, would have been a cumbersome and unreliable process and would have led to results with low validity. Therefore, we came to the idea that by analyzing an as full as possible body of related texts from the time period covered would provide a usable source. Such texts, by applying text mining should cover all the information that is needed for constructing reasonably reliable historical time series, which would allow for the extraction of knowledge even from unstructured or poorly structured text files. The collection and representation of textual knowledge and information from these various document sources was using machine intelligence. The assumption was that this tool could be used to generate a historical time series for the description of waste management systems.

In this stage, a collection of texts of nearly 115 directives, regulations and strategic documents was examined. Thereafter, the word frequencies distributed over the decades under examination could be determined and so, after the classification of the potential factors and sub-factors into 33 subsystem elements, frequency tables for the examined decades and subsystem elements were produced, in which the role of each word or expression was determined by its frequency. The purpose of this research phase was therefore to produce a time series for each newly proposed factor based on the thus explicitly identified hidden historical data.

It is obvious that due to its complexity, the 33-factor model should be much more accurate in describing the system components than the simple and rough six-element model used in the literature and the first phase of our research. Although the complex model was created by decomposing the six factors into sub-elements, now we assumed that these are independent concepts (factors). After analyzing the basic and detailed model and the relationship matrices, it was discovered that the two models were conceptually different and thus the new, finer model could be used as the starting point. However, such a complicated FCM and its graph are beyond the comprehension and cognitive analysis abilities of any human expert and we assumed that by reducing the model again—however, in a way where the adequacy of the components and the interconnections is kept, one or several completely novel models for the RWMS could be obtained. The goal was to have an adequate but still interpretable model, which could be later used by the analysis of any local RWMS.

2.3 Sustainable Waste Management Systems—A Necessity

The current situation in municipal waste management has been characterized by a major transformation in recent years. This transformation from the wider environment is driven by the new development policy guidelines along the lines of the global and European economic environment and sustainability, which aim to create a new, greener, more material-and energy-efficient "European resource-efficient, recycling society". All this helps to create a livelier, healthier environment. In the future, environmentally friendly waste management, the use of environmentally friendly and innovative technologies will be emphasized, and the use of hazardous materials will be reduced. Waste must be viewed by society as a resource. Landfill decreases; in the future, only non-recoverable waste will be disposed of, and the amount of biodegradable waste will be minimized.

For these goals to be met in a timely manner, a complete overhaul of domestic waste management is needed; systems must be established, maintained and operated, the elements of which can ensure the long-term efficiency and development of domestic waste management, thereby achieving the relevant policy objectives.

The characteristics of a complex waste management system are that it solves the waste management of the territorial unit in a geographically connected area, covering all the settlements of the area, according to the same environmental and technical aspects, according to the same sorting principles and at the same technical level.

Regional waste management systems must be economically affordable, environmentally effective and socially acceptable. This includes, among other things, practical elements of waste management (e.g. transport, treatment and disposal) and people's attitudes towards the topic (how they think about selective collection, recovery, incineration, etc.) [13].

A municipal association established in a given region should optimally design the selective collection of raw materials and bio-waste and of the safe disposal of the landfill, in accordance with the geographical, economic and social environment in which it operates. This requires the development of waste collection points, waste yards, transfer stations and a waste sorting facility to be implemented in a complex waste management centre, mechanical and biological waste management technologies, as well as modern and safe landfill facilities and logistical and technological facilities and machines.

The system of regional treatment centers will help to increase the level of mechanization, operate the deposition technology more efficiently and economically, and use the resulting biogas for energy purposes; it also facilitates the use of complex waste treatment processes (composting of green and bio-waste, preparation of selectively collected components for industrial use, pre-treatment of construction waste for recovery). The development of regional systems necessarily entails the need to change the collection and transport system.

The expected results of the regional waste management system as a result of the above are the following: less waste going to landfill, more waste recycled, less material to be disposed of in incinerators, reduced environmental load and demand, more efficient and more recyclable waste treatment, reduced environmental pollution from illegal or unauthorized disposal and environmental awareness is expected to increase.

2.4 Current Evaluation Methods

The essence of evaluating interventions is to judge the value of an intervention by specifically gathered and analyzed information according to explicit criteria. The purpose of evaluation is partly to reinforce the repeatability of success actions, and partly to avoid the repetition of mistakes, and partly to report to the citizens. The goal is to analyse success and failure factors so that promoters can adjust their own intervention and program implementation work.

Evaluation is also a tool for strategic and operational decision-making. The practical benefit is to link the objectives, resources, results and impacts of the intervention to increase the efficiency and effectiveness of the project or development. In Europe, evaluation has begun to develop in some Northwest European countries (England, Sweden, Denmark, the Netherlands), where standards are now in effect. However, by now, each EU member state has adapted the assessment with some elements of its own traditions and is now preparing its deployment.

When assessing the sustainability of regional development, it is, first of all, necessary to distinguish between the concepts 'assessment for sustainability' and 'sustainability assessment'. The former examines the process in question and determines the route to be taken to assess sustainability, whereas the latter is a static report.

The results published in the international literature make it clear that the interest of researchers in developing and applying new methods is still being watched. These methods are basically aimed at decision support systems (DSS) and help decision-makers to choose the optimal option from among the possible alternatives (recovery, incineration, disposal).

In order for the systems being analyzed, it is necessary to define the system boundaries that clarifies which elements do belong to the system under investigation and which are outside its boundaries. According to [5], every technical system is influenced by factors outside the system's boundaries, but it also influences the system's own environment. This interaction between the system and its environment can be characterized by input and output parameters.

According to the research [1], the system is created during the accumulation of sub-units and their interrelations, where the "whole" (unit) is more than the mere sum of the parts. Therefore, in their investigations, they focused their attention on the components and the relationships among them. Thus, by their interpretation of the "whole", it is more complete than the whole set of components. In addition, several other studies (e.g., [7, 9]) dealt with the system theory approach of integrating components into a single system, which provides a conceptual framework for the problem and addresses the system by: the components interconnected to form a higher level unified system.

In order to be able to objectively evaluate the sustainability of any type of infrastructure system, we need to consider its complexity when making decisions. This requires a comprehensive system science approach, as the technical, social and environmental systems are closely linked. The journey to sustainability is called a "demonic" challenge as it may include complexity, uncertainty, multiple perspectives, and multiple stakeholders, competing values and even ambiguous terminology. Thus, there is an urgent need for a better understanding of dynamic complex systems in general, and infrastructure systems in particular.

Thinking and acting in a systemic way is often not easy, but it must become an essential part of our way of life if we are to develop the world in a sustainable way. To master systemic thinking, it is necessary to move away from our everyday attitudes.

In the case of RWMS as an open system, it is important to define the system boundaries in order to evaluate the relationships between the components [10]. A thorough literature review of the system approach to waste management has been published in [5]. The purpose of analyzing waste management systems is to gain a comprehensive picture of the impact of different technological options and to understand the environmental, economic and social impulses. In this way, system analysis provides policy makers with an opportunity to support decision-making towards sustainability based on the information available. It is worth developing and applying a computer program to enhance this process.

According to the research of Schneller et al. [8], complex systems are those systems in which several properties are considered simultaneously and the simultaneous and combined evaluation of properties is a problem, i.e. it is not a trivial task (cf. also Kóczy et al. [6]).

Kiss [4] articulates the systemic relationship between man and his environment: as our species becomes more and more clustered in resources and flows in material and energy, our responsibility for maintaining planetary equilibrium increases. Each time we make a significant change to any part of the natural regulatory process, we increase the likelihood that one of the changes that will occur will affect the stability of the entire system. This is because any operating system seeks to adjust and search for optimum deviations from its steady state due to changes in energy flow or response time. In order to provide feedback to the process, it needs to be decided within a short time to what extent and how the by-products released can be traced back to the primary processes.

The design of RWMSs is fundamentally influenced by natural and constructed environmental factors. However, the relationship is two-sided, since not only the waste management system affects the environment, but also the environment affects the system under investigation. There is a continuous flow of material, energy and information between the RWMS and its built and natural environment. We know relatively little about their interactions with each other, especially in a system approach.

A complex examination of the RWMS and its environment enables the detection of previously unexplored or insufficiently investigated interactions that fundamentally determine the operation of the system and its sustainability, and determine future development opportunities.

The diversity of integrated systems related to environmental engineering has sparked a number of initiatives to develop new methodological tools for controlling current conditions, assessing future risks and presenting possible future perspectives.

The elements of these systems are connected by material, energy and information flows and form a coherent network. One of the definitive elements of waste management systems are the people (individuals or organized social units) themselves, who can be, for example, waste producers, tax payers, environmental sufferers, etc. The information flow in a broad sense plays an important role in organizing the relationships between humans and the built and natural environment. Integrated municipal waste management systems are therefore designed and operated on a spatial and temporal scale.

For policymakers, understanding complex systems is the key to success, but it also carries significant risks. Complex problems require careful analysis. Below a certain level of complexity, the system can be satisfactorily modeled by analytical tools in mathematics, but in the case of really complex systems, we have to look for other methods. In such cases, computer simulation and the application of the tool box of Computational Intelligence (CI) can be a useful additional tool to study the problem in a cost-effective and safe manner.

In such complex systems, human behaviour can be understood as a reaction to the interpretation of signals from the environment. Commitment to a given case is based on individual and group cumulative knowledge. Different forms of knowledge reflect different forms of commitment to the environment and provide different perspectives on the same complex system. In RWMSs, stakeholder knowledge can play an important role. Stakeholder knowledge can be grouped into traditional knowledge, local knowledge and expertise [12].

One possible solution to an integrated approach to environmental planning is the direct involvement of stakeholders. This approach requires an approach that facilitates communication between multidisciplinary teams and experts. There is a need for a participatory management framework that offers a meaningful change in the exploitation of local knowledge. With the involvement of stakeholders, participatory management strengthens strategic focus, reduces uncertainty, improves control and increases implementation efficiency.

Knowledge representation addresses the fundamental issues of complex knowledge-based systems. In doing so, experts are asked to transfer their knowledge and expertise so that the conclusions of the study are realistic [3]. An integrated municipal waste management system is a complex and extensive system where the system needs to be implemented and operated in a sustainable manner, taking into account the needs and needs of different experts and stakeholders.

2.5 Summary of the Problem Context

For large and complex systems it is extremely difficult to describe the entire system by a precise mathematical model [2]. RWMSs are real and important elements of everyday life, therefore, problems generated from these systems are real problems when looking at a solution. The new system model outlined in the research described in the first part of this volume is of such a quality that it can have a significant impact on solving environmental problems across systems. The system as a whole can have an impact on complex rural development, the cultivated, nature-friendly, local society, essential to the future, and its relationship with sustainable development.

As it was indicated, we proposed the application of the principles of the system-of-systems approach to break down the RWMS model into subsystems. In the following chapters, it will be presented how the modeling procedure went step by step, first by the process of applying a model based on the six major system components in the literature, and then by applying the results of the expert workshop for an extension of the size and complexity of the model, and by determining hidden historical data by the subsequent text mining technique, and finally, by developing a new FCM reduction algorithm and thus transforming the new model into various interpretable ones by applying the new algorithm under various assumptions and conditions.

Acknowledgements The paper was written with the support of the project titled "Internationalisation, initiatives to establish a new source of researchers and graduates and development of knowledge and technological transfer as instruments of intelligent specialisations at Széchenyi István University" (project number: EFOP-3.6.1-16-2016-00017).

References

1. Boardman J, Sauser B (2006) System of systems—the meaning of of. In: 2006 IEEE/SMC international conference on system of systems engineering. IEEE, p 6
2. Jamshidi M (2008) Systems of systems engineering: principles and applications. CRC Press
3. Ketipi MK, Koulouriotis DE, Karakasis EG, Papakostas GA, Tourassis VD (2012) A flexible nonlinear approach to represent cause–effect relationships in FCMS. Appl Soft Comput 12(12):3757–3770
4. Kiss T (2007) Economics and system approach in municipal solid waste management. University of Miskolc, Faculty of Engineering Earth Sciences, Institute of Raw Materials Preparation and Environmental Procedures
5. Klang AB (2005) Sustainable Waste Management. Methods and Framework for analysis, Fakultet for ingeniørvitenskap og teknologi
6. Kóczy LT, Medina J, Reformat M, Wong KW, Yoon JH (2019) Computational intelligence in modeling complex systems and solving complex problems. Complexity 2019:Article ID 7606715
7. Mittal S, Zeigler BP, Martin JLR, Sahin F, Jamshidi M (2008) Modeling and simulation for systems of systems engineering. Systems of systems–innovations for the 21st century
8. Schneller K, Podmaniczky B, Podmaniczky L (2005) Application of complex systems comparison method in agricultural spatial function analysis. Acta Agraria Kaposváriensis 10(3):259–278
9. Sterman JD (2001) System dynamics modeling: tools for learning in a complex world. California Manag Rev 43(4):8–25
10. Sundberg J, Gipperth P, Wene C-O (1994) A systems approach to municipal solid waste management: a pilot study of göteborg. Waste Manag Res 12(1):73–91
11. Torma A, Hatwágner FM, Buruzs A, Kóczy TL, Susniene D (2014) Developing a new method to model the causal relationships in stakeholder management at companies by using fuzzy cognitive map approach. In: Kóczy T, Tormási A, Kulczycki P, Talaba D, Kacprzyk J (eds), Proceedings of the 6th Győr symposium and 3rd Hungarian-Polish and 1st Hungarian-Romanian joint conference on computational intelligence. Győr, Széchenyi István University, pp 211–215
12. Wildenberg M, Bachhofer M, Adamescu M, De Blust G, Diaz-Delgadod R, Isak KGQ, Skov F, Riku V (2010) Linking thoughts to flows-fuzzy cognitive mapping as tool for integrated landscape modelling. In: Proceedings of the 2010 international conference on integrative landscape modelling and linking environmental, social and computer sciences, Montpellier, Feb 3–5. Citeseer, 2010
13. Wilson EJ, McDougall FR, Willmore J (2001) Euro-trash: searching Europe for a more sustainable approach to waste management. Res Conserv Recycl 31(4):327–346

Chapter 3
Identification of the Main Factors (Concepts) of Waste Management for Building the Initial FCM Model

Adrienn Buruzs

3.1 Introduction

When continuing the presentation of how to build the initial FCM model in the case study, in the next, the identification of the main factors of the system under investigation, formally, of the concepts of the FCM will be done. The idea of 'main driving forces' can be defined as the factors that change the status quo of an operating waste management system for good or bad, be it any technological or human aspect, or even, legislation that encourages an integrated waste management approach, or the change of public perception of the issue.

There is a large amount of technical papers in the related literature describing the factors affecting the operation and sustainability of the class of systems on hand, namely, municipal waste management systems. As methods for municipal waste management systems have evolved, also the number of factors recognized as affecting the system has increased dramatically worldwide.

The so-called horizontal factors describe the processes of change in the generation of different types of waste (the shift between municipal waste, slag waste, recyclable waste and illegally disposed waste). Vertical factors represent changes in the total amount of waste streams as a result of demographic, economic, social and technical developments (quantitative and monetary data) [1].

Based on the literature review, the authors conclude that there is strong consensus in the relevant literature that the typical RWMS contains at least the following six major factors: *environmental*, *economic*, *social*, *institutional*, *legal*, and *technical*. These factors are the 'main drivers' of sustainable RWMS (Table 3.1), which determine why the system under study operates in the observed way [2, 7–11]. Initially, the definition of the main factors found in the literature has been accepted by the authors. However, during the modelling procedure and analysing some of the results, it became evident that to redefining the factors and subdivide them into subsystems at a later stage of the research is unavoidable and necessary if an adequate model is constructed.

L. T. Kóczy, *Fuzzy Cognitive Maps*, Studies in Fuzziness and Soft Computing 427,
https://doi.org/10.1007/978-3-031-37959-8_3

Table 3.1 'Key Drivers' of IWMS and their respective subsystems

Factors	Subsystem elements
Environmental factors	Emissions; Climate change; Land use; Recovery and recycling targets; Depletion of natural resources; Human toxicity
Economic factors	Efficiency at subsystem level; Efficiency at system level; Available funding/subsidies; Equity; System costs and revenues; Pricing system for waste services, Secondary materials market
Social factors	Public opinion; Public participation in the decision making process; Risk perception; Employment; Local demographics—population density, household size and household income; Public resistance (NIMBY—Not In My BackYard, LULU—Locally Unacceptable Land Use)
Institutional factors	Local and regional politics and planning; Managerial conditions and future directions; Institutional and administrative structure of waste management
Legal factors	Relevant legislation (international, national, regional and municipal)
Technical factors	Collection and transfer system; Treatment technologies; Waste stream composition and change

3.2 The Methodological Approach of Accumulating Information

The method of the primary collection of information on RWMSs was the use of questionnaires presented to a pool of RWMS experts. Questionnaire surveys belong to the family of quantitative methods, whose primary purpose is to accumulate (large) amounts of data in a closed form. "Closed form" in this context means that the questions are fixed and are applied unchanged to all interviewees [4].

Only a few of the benefits of online questionnaires were directly used in this case study research (they are considered flexible, fast, cheap, and available in large numbers without geographical restrictions), and not all the potential disadvantages (coverage, need for expertise, required hardware capacity, special populations) were significantly manifesting here. Expertise was a prerequisite, as only experts were sought out; the availability of the necessary hardware capacity was not an issue either; and choosing a special base population was no problem as there was a large address list.

The main argument for applying online research was cost effectiveness. Online questionnaires need much less money (they are almost free), are much quicker than questioning persons in the frame of personal meetings, and so, much more respondents can receive and answer the questions. As Kosko [6] pointed it out, FCM reliability increases when the number of the expert samples increases which serve as input data when setting up the model. Another important aspect when designing the online questionnaires was their easy and simple completion. This aspect guided the researcher team's hand in formulating the questions themselves [5].

In the next paragraph, a few further remarks about the formulation of the questionnaires are given. It should be taken into consideration that analysing data, as in the case of other methods as well, it is necessary to refine the data and adjust the weighting of the individual questions [3]. An important principle when it comes to questions is clear wording. As the researcher already has a deep understanding of the topic, they may fall in the trap of asking complex questions clearly understandable to them, but far from having an unambiguous meaning to the respondents. Simplicity and brevity should be sought always when formulating questions. When editing the questionnaire, it is advisable to ask closed questions, in order to achieve the desired result.

3.3 Results of the Questionnaire Survey on RWMSs

It should be noted that the review of the situation and the sustainability assessment of RWMSs require an interdisciplinary approach, since the topic must be examined from the social, economic, environmental and legal, etc. aspects. The aim of applying the questionnaire method was to create an as objective as possible, and at the same time, a somewhat innovative tool that could further support decisions about the sustainability of waste management.

As mentioned above, the researcher team chose the method of online questionnaire survey, where the main purpose of the questionnaires was to identify the impact of each of the determining factors of waste management systems and the extent of the impact, based on the feedback from professional experts. It should be, however, noted that expert responses are unavoidably subjective and contain elements of uncertainty. This is why the model to be created was chosen as a fuzzy model. So, based on the responses, it had to be examined first how the factors affecting the sustainability of the RWM systems interact with each other, according to the respondents' opinion.

After presenting the six factors under discussion, respondents were asked to rate the extent to which one factor affects the others, on a scale of -5 to $+5$. (Applying a kind of signed Likert scale approach.) The team requested that the causal relationships between the factors be evaluated, and their directions and strengths be determined by this predefined scale capable of describing any relationship between an ordered pair of factors, whether it was positive, negative or neutral. Thus, in the process of the online questionnaire survey, the respondents directly assessed the relationships between the main factors of the waste management system according to their most educated expert opinions. In the subsequent modelling phase, the views of each respondent were taken into account. The responses received were put in tables, and answers to pre-coded closed questions were processed by simple summaries.

The strength of the interaction between the factors was calculated from the median value of the questionnaire responses. Subsequently, calculations of the fuzzy graph structure model were made, to producing a number of variants, allowing the evaluation of each factor to be more or less independent of the subjective judgment of individual experts.

Table 3.2 shows the median values for the cumulative responses for each topic. The factors of the matrix are as follows:

Table 3.2 Medians of data from the questionnaire

	C_1	C_2	C_3	C_4	C_5	C_6
C_1	0	4	4	4	2	2
C_2	3	0	3	3	2	2
C_3	4	3	0	3	2	2
C_4	2	3	2	0	2	2
C_5	3	3	2	3	0	3
C_6	2	2	2	2	2	0

- C_1: technical, technological factor;
- C_2: environmental factor;
- C_3: economic factor;
- C_4: social factor;
- C_5: legal factor;
- C_6: institutional factor.

It is an interesting fact that, as it can be seen from the above connection matrix, the median value of all responses received represented only positive inter-factor relationships, but it should be noted that, in the individual responses, some practitioners assigned inter-factor effects on the whole scale ($-5 +$ up to 5), including negative values.

A clear problem of this modelling approach is that all the six factors are extremely broad concepts, and that the subsystem elements identified within each factor may have different effects on the other factors, not to speak about the other subsystem concepts. This raised the question whether the respondents understood the questionnaire in the same or at least in similar ways. The answer to this very basic question is that in this case, the concepts could not be accurately delineated, since when evaluating each factor, it could be assumed that the respondents did not associate with exactly the same concepts and this could have significant influence on the answers, causing a divergence of the Likert scale values assigned to the same ordered pair by different experts. This uncertainty was further increased by the fact that the questionnaires were completed online, so there was no way to explain the concepts and the method in sufficient detail to the participants of the survey.

3.4 Applying the Results of the Questionnaire Survey

As mentioned several times previously, the model contains two input data sets. One is the expert database, also known as the connection matrix, which is based on the experience and knowledge of the experts. The values in the responses to the questions were transformed by a linear mapping to the interval $[-1;\ 1]$, thus obtaining a signed fuzzy membership degree for each ordered pair of factors, and the medians of these

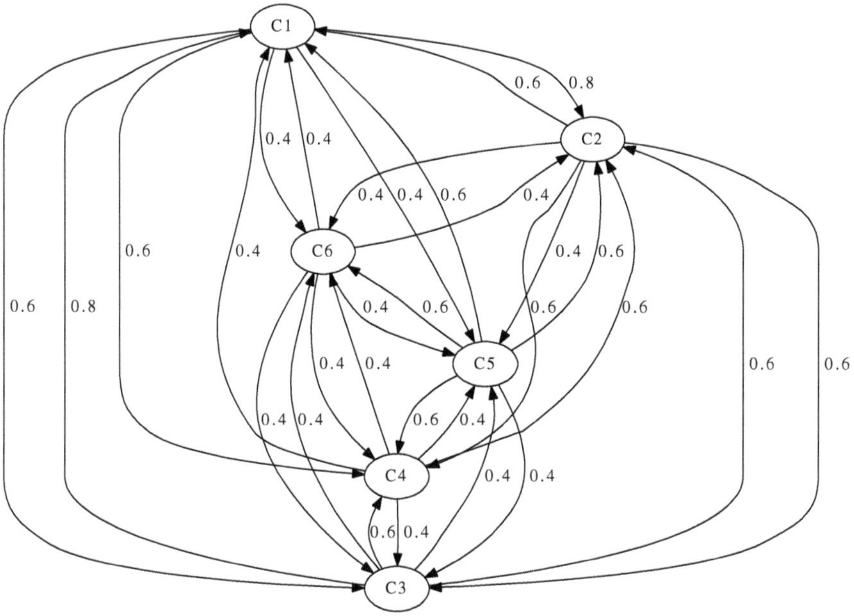

Fig. 3.1 Initial fuzzy cognitive map

values extracted from the $N = 75$ responses were then used as the input data of the fuzzy cognitive map modelling the RWMSs (see Fig. 3.1). This way, the first version of the connection matrix of the simulation model (Table 3.3) was completed. This connection matrix contained the factors and the relationships between them, with their directions and strengths, representing the typical opinion of a domain expert.

In the graph in Fig. 3.1 and in the Table 3.3

- C_1: technical, technological factor
- C_2: environmental factor
- C_3: economic factor
- C_4: social factor

Table 3.3 Initial connection matrix

	C_1	C_2	C_3	C_4	C_5	C_6
C_1	0	0.8	0.6	0.6	0.4	0.4
C_2	0.6	0	0.6	0.6	0.4	0.4
C_3	0.8	0.6	0	0.6	0.4	0.4
C_4	0.4	0.6	0.4	0	0.4	0.4
C_5	0.6	0.6	0.4	0.6	0	0.6
C_6	0.4	0.4	0.4	0.4	0.4	0

- C_5: legal factor
- C_6: institutional factor

assigned.

Note that the graph contains no selfloops, and the diagonal of the connection matrix only contains 0-s, which is in accordance with the Type 1 FCM structure.

3.5 Summary

As the first step in the case study research, the researcher team reviewed the relevant domestic and international literature in order to establish a starting FCM model. Based on the literature, six key factors of municipal waste management were identified, which are generally accepted in the literature. There is strong consensus in the relevant literature that the typical RWMS contains (at least) these six major factors. Thus, these factors are the 'main drivers' of any (sustainable) RWMS, and they determine why the system under investigation operates in one or another way. In order to refine the model, the 'knowledge base' thus acquired was complemented by additional information obtained through scientific discussions at domestic and international conferences in the field.

In order to gain expert knowledge, the team chose the method of an online questionnaire survey. The purpose of the questions was to identify the impact of each of the determining factors of waste management systems and the intensity of these impacts, based on feedback obtained from professionals. Subsequently, the FCM graph structure model was determined, based on calculations, which allowed the evaluation of the effect of each factor to be estimated independently of subjective judgment.

Acknowledgements The paper was written with the support of the project titled 'Internationalisation, initiatives to establish a new source of researchers and graduates and development of knowledge and technological transfer as instruments of intelligent specialisations at Széchenyi István University' (project number: EFOP-3.6.1-16-2016-00017).

References

1. Beigl Peter, Lebersorger Sandra, Salhofer Stefan (2008) Modelling municipal solid waste generation: A review. Waste management 28(1):200–214
2. Den Boer J, Den Boer E, Jager J (2007) Lca-iwm: a decision support tool for sustainability assessment of waste management systems. Waste management 27(8):1032–1045
3. Eranus E, Láng S, Máth A, Rácz A (2005) New methods of questionnaire data collection: telephone, computer assisted (capi, cati) and internet data collection. In: Letenyei L (ed) Settlement research, methodological manual and text collection. TeTT Books
4. Kmetty Z (2011) Research methodology and presentation. Part 4: questionnaire design in theory and practice. Károli Gáspár University of the Reformed Church, TÁMOP-4.1.2.A/1-11/1-2011-0091 'Information-Knowledge-Validation'

5. Kontra J (2011) Methodology of pedagogical research. University of Kaposvár
6. Kosko Bart (1988) Hidden patterns in combined and adaptive knowledge networks. Int J Approx Reason 2(4):377–393
7. Lang DJ , Binder CR, Stauffacher M, Ziegler C, Schleiss K, Scholz RW (2006) Material and money flows as a means for industry analysis of recycling schemes: a case study of regional bio-waste management. Res Conserv Recycl 49(2):159–190
8. Morrissey AJ, Browne J (2004) Waste management models and their application to sustainable waste management. Waste Manag 24(3):297–308
9. Thorneloe SA, Weitz K, Barlaz M, Ham RK (1999) Tools for determining sustainable waste management through application of life-cycle assessment: update on us research. In: Seventh international waste management and landfill symposium V, pp 629–636
10. Van de Klundert Arnold, Anschutz Justine (1999) Integrated sustainable waste management: the selection of appropriate technologies and the design of sustainable systems is not (only) a technological issue. CEDARE/IETC Inter-Reg workshop Technol Sustain Waste Manag 13:1–17
11. Wilson EJ, McDougall FR, Willmore J (2001) Euro-trash: searching Europe for a more sustainable approach to waste management. Res Conserv Recycl 31(4):327–346

Chapter 4
Rethinking the Set of Concepts—Applying Expert Analysis by the Workshop Method

Adrienn Buruzs

4.1 Introduction

The development process of the RWMS model requires a significant social analysis, especially, the examination of the technical features (such as collection, transportation, equipment, etc.), the analysis of legal and institutional issues based on personal experience, and the synthesis of possibly conflicting expert opinions. Although personal experience and expertise are generally considered subjective, their contribution to the design of the waste management system is crucial.

It seemed the only way to get a more refined and more adequate FCM model of the RWMS structure was to entirely rethink the set of potential composing factors, i.e., the concepts of a new FCM, and to determine the values assigned to each directed edge, i.e., each degree and sign of the pairwise influence effects. In order to do so, a workshop of experts was organised where all main areas represented in the original six-concept model were represented in a sufficient number, to be able to determine the concepts and, to some extent, the mutual effects on each other.

When discussing this phase of the case study, the authors intend to present a more generally applicable and efficient methodological approach for creating the FCM corresponding to a complex multicomponent system where there is no objective way to set up the model, however, sufficient amounts of expert domain knowledge is available. In the workshop, the ways of perception and assessment of the structure and components of such complex systems conceited by an expert panel were also revealed, especially, by the analysis of future RWMSs, and their perspectives and risks in the light of economic, social, environmental, legal, institutional and economic factors. The approach presented here may be applied to a wider range of similar systems in many fields of applied sciences.

L. T. Kóczy, *Fuzzy Cognitive Maps*, Studies in Fuzziness and Soft Computing 427,
https://doi.org/10.1007/978-3-031-37959-8_4

4.2 The Methodological Approach—the Workshop Based Analysis

We decided that when organising the workshop, it would be helpful to establish a systematic formal guide that describes each step in the design of the new FCM. This procedure could also serve as a general guideline for workshop operation when for other similar problems, a refinement of the initial FCM model is targeted [3]. The methodological framework of the application and development phase is described below.

In the present case, the process involved four steps:

1. As it was explained in the previous Chapter, based on the literature, the authors established the six main factors (concepts) of RWMSs, and presented them to the expert group;
2. The workshop participants were sub-divided into smaller groups of specialists who then subdivided each factor into sub-factors (refined new concepts);
3. The specialist sub-groups also assessed the directions and extents of all interactions between the pairs of these factors; both these new concepts and the values assigned to the interactions were presented to the entire workshop;
4. Based on the above, the new refined FCM was created as the starting input of the simulation.

The essence of the workshop methodology is a structured, consciously organized and guided group activity in which participants effectively solve tasks and collect diverse solutions using a variety of tools and techniques. The purpose of this technique, beyond the solution, is to generate as many good alternative solution ideas as possible, so that there is good collaboration within the group, regardless of the personalities of the participants, and thus the experts may harvest the benefits (synergies) from working together [1].

Interactive activities should be planned consciously during the workshop, as a large number of experts may be affected and complex organizational and logistical measures may be required. Therefore, briefing and planning before a meeting can be crucial to the success or failure of a job.

During this briefing and planning phase, the promoters of the event have a key role to play, where the following checklist being helpful:

- Definition of the goal of the workshop (Case study: Determining the refined FCM model for the given RWMS problem)
- Identification of the topics and specific issues to be discussed (Case study: Determining the concepts (factors) and the pairwise mutual interaction signs and degrees
- Compilation of the list of external experts to be invited (Case study: experts of environmental issues of all six main areas in the initial FCM)
- Technical issues, like providing the venue (possibly with catering), listing the preliminary program, schedule and possible outcomes of the work, etc.
- Selection of the working method and the moderation method (Case study: Sub-groups of specialists in separate sessions, plenary presentations of the sub-groups)

- Communication strategy, definition of dissemination of results (Case study: transfer of the resulting model to the modelling and IT specialists)

The workshop method is both time consuming and time saving. It takes time because it takes serious organizational preparation in every way, and the two-level sessions must be serially organised. At the same time, it is time-saving because the intensive work of the participants can lead to drastic changes and innovations in the outcome of the work by mobilizing the energy and expertise of the various stakeholders (sub-topic specialists), in this case study, all sub-topics within the RWMS.

During the workshop, the moderator emphasizes the complexity and essence of the fuzzy cognitive map; for example, by limiting the number of (sub)system elements to be developed, or by focusing participants' attention on a few narrower topics. However, this requires the moderator to be familiar with the basic concepts and context of system modelling. During the process, it is important for the moderator to use appropriate methodological tools in order to gain the most effective knowledge of the participants on the topic. It is advisable for the moderator to create his/her own FCM on the topic before starting the workshop. Like many participatory approaches, the moderator plays a crucial role in the design process [2, 4].

At the workshop the work schedule was as follows:

1. presentation of the topic (theoretical background and working method),
2. defining the subsystems of the main factors by small group work,
3. presentation of the results of small group work,
4. identification of subsystem links and determination of link strength.

The objectives of the workshop were explained to the participants:

- Mapping the RWMS network;
- Determining the impact of factors influencing the sustainability of RWMS;
- Definition of the subsystems of the dominant factors of the RWMS, the relationships between the system components and the strength of these relationships;
- Using the results of collaborative work as input data for a soft computing method for RWMS modelling.

Before the start, we set out the framework (duration) and the working method of the workshop. For the latter, we chose the method of frontal and interactive workshops:

1. in the framework of the frontal method, we introduced the key issue of the work, the current phase and methodology of the research, and its intended goal;
2. the interactive approach served to systematically formulate and present the participants' suggestions.

In this particular case study, the breakdown of the main components into subsystems was carried out by a team of twelve experts. At the beginning of the work, the authors described the purpose of the research, the driving forces identified from the literature, the essence of the FCM method and the mapping process, and the planned phases of the workshop. The moderator was the main contributor. She was supported by two invited colleagues in defining the professional framework.

Subsequently, the experts were asked to form six groups of two persons each. We have taken into account their professional knowledge and work experience; based on this, we found that two participants came from the area of operation (technical factor), the legal area (legal factor), the authorities (institutional factor) and economic area (economic factor). The staff of the Department of Environmental Engineering formed groups representing social and environmental factors.

During the work in small groups, participants were asked to identify subsystem elements in their particular area of expertise. To avoid two much fragmentation, experts were asked to identify up to 5–7 subsystem elements per component. In the course of their work, the team members identified the factors that they considered important to the operation of the system, given their professional experience and knowledge. (As a matter of course, in a real life situation, with more funding, a larger number of specialists with larger subgroups can be invited.)

We assisted the work of the small groups by asking the following questions:

- What are the subsystem elements of the RWMS that can be identified? What factors, things do you remember when RWMS is mentioned to you?
- How to interpret the structure and role of each sub-factor, e.g. in respect to the main factor?
- How do experts perceive the impact of individual factors on other (sub)factors? Are there positive or negative effects? To what extent does factor *A* influence factor *B*?

4.3 Results of the Questionnaire Survey

Defining the properties of the subsystem elements is very important because certain criteria must be met. When creating an FCM, it is important to consider that, in order to determine the effect of another factor on the parameter being investigated, the factors must be quantifiable. It is important to understand how FCM works and to understand the map so that the factors are clearly defined [5]. Table 4.1 describes the subsystem elements for each of the six components.

The FCM approach allowed participants to identify potential conflicts in the discussed models and propose solutions for such conflicts. Small group sessions helped to understand common or divergent priorities among the disciplines. The FCM approach required that participants reflect on how each element influenced each other within the system under study. As the FCM approach focuses on causal relationships between factors, it naturally directs common thinking toward a problem-solving approach.

Table 4.1 The identified sub-factors of the main factors and the concept IDs (CID) of them

Main factor	Sub-factor	CID
Technology (C_1)	Engineering knowledge	$C_{1.1}$
	Technological system and its coherence	$C_{1.2}$
	Local geographical and infrastructural conditions	$C_{1.3}$
	Technical requirements in the EU and national policy	$C_{1.4}$
	Technical level of equipment	$C_{1.5}$
Environment (C_2)	Impact on environmental elements	$C_{2.1}$
	Waste recovery	$C_{2.2}$
	Geographical factor	$C_{2.3}$
	Resource use	$C_{2.4}$
	Wildlife (social acceptance)	$C_{2.5}$
	Environmental feedback	$C_{2.6}$
Economy (C_3)	Composition and income level of the population	$C_{3.1}$
	Changes in public service fees	$C_{3.2}$
	Depreciation and resource development	$C_{3.3}$
	Economic interest of operators	$C_{3.4}$
	Financing	$C_{3.5}$
	Structure of industry	$C_{3.6}$
Society (C_4)	Public opinion	$C_{4.1}$
	Public health	$C_{4.2}$
	Political and power factors	$C_{4.3}$
	Education	$C_{4.4}$
	Culture	$C_{4.5}$
	Social environment	$C_{4.6}$
	Employment	$C_{4.7}$
Law (C_5)	Monitoring and sanctioning	$C_{5.1}$
	Internal and external legal coherence (domestic law)	$C_{5.2}$
	General waste management regulation in the EU	$C_{5.3}$
	Policy strategy and method of implementation	$C_{5.4}$
Institution (C_6)	Publicity, transparency (data management)	$C_{6.1}$
	Elimination of duplicate authority	$C_{6.2}$
	Fast and flexible administration	$C_{6.3}$
	Cooperation among institutions	$C_{6.4}$
	Improvement of professional standards	$C_{6.5}$

4.3.1 The New Refined FCM Model

As the result of the workshop and system approach, we replaced the former too simple six-element model with the new, much more complex model of RWMS. Figure 4.1 illustrates the 33-factor system map obtained from the expert analysis of the subsystem elements, based on the original major system components.

In the next step, the types and values of causal relationships between the subsystem elements were identified by the plenary expert panel, where the weights of the relationships between the elements were also determined using the scale $\{-5, \ldots, +5\}$ (with 11 values). These values were then linearly transformed into signed fuzzy membership degrees.

In this way, a relationship matrix of 33 elements was created which contained the main factors, the subsystem elements and the respective relationships between each pair of them. This relationship matrix contains 1056 (33×32, $+/-$ and 0) relationships (Tables 4.2 and 4.3).

Although the graph would only need to display non-zero relationships, which is less than 1056, the non-zero elements were still too many for producing an interpretable and clearly comprehensible graph.

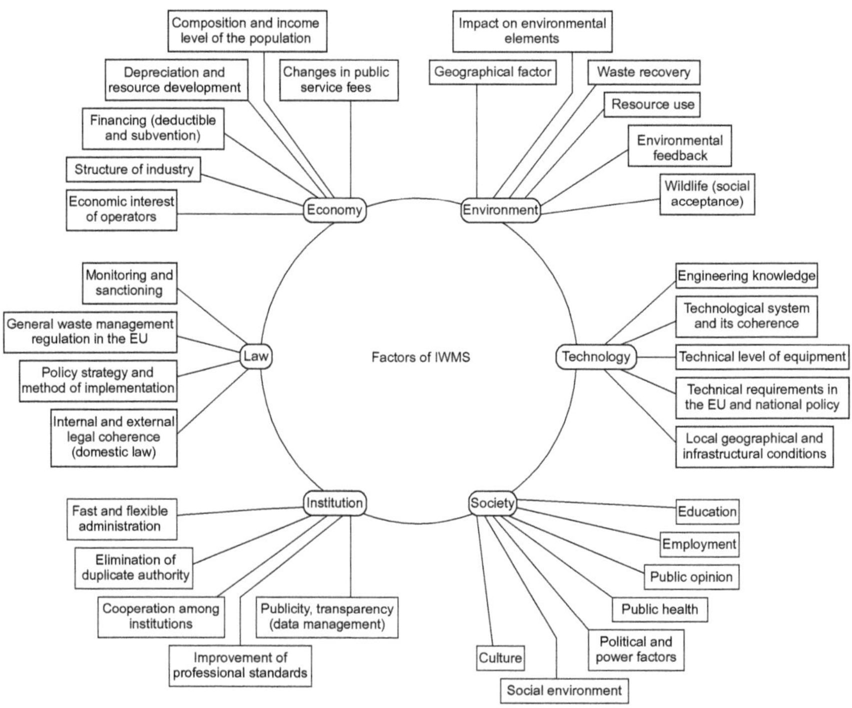

Fig. 4.1 The Thirty-tree Sub-factors of the IWMS Model

Table 4.2 Refined connection matrix created by experts as a result of the workshop, Part 1

CID	$C_{1.1}$	$C_{1.2}$	$C_{1.3}$	$C_{1.4}$	$C_{1.5}$	$C_{2.1}$	$C_{2.2}$	$C_{2.3}$	$C_{2.4}$	$C_{2.5}$	$C_{2.6}$	$C_{3.1}$	$C_{3.2}$	$C_{3.3}$	$C_{3.4}$	$C_{3.5}$	$C_{3.6}$
$C_{1.1}$	0	0.2	0	0.6	0.4	0.6	0.2	0	0.8	0.2	0.6	0.4	0.8	0.4	0.8	0.4	0.4
$C_{1.2}$	0.4	0	0.4	0.4	0.6	0.2	0.2	0	0.4	0.2	0.4	0.6	0.8	0.6	0.6	0.6	0.6
$C_{1.3}$	0	0.2	0	0.2	0	0	0	0	0.2	0	0.4	0.6	0.6	0.6	0.6	0.4	0.4
$C_{1.4}$	0.6	0.4	0.2	0	0.8	0.2	0	0	0	0.6	0.6	0	0	0.2	0.2	0	0
$C_{1.5}$	0.8	0.2	0	0.8	0	0.4	0.2	0	0.4	0.4	0.6	0.6	0.8	0.6	0.6	0.6	0.6
$C_{2.1}$	0	0	0.6	0.2	0	0	0	0	0.2	0.4	−0.6	0	0.2	0	0	0	0
$C_{2.2}$	0	0.2	0	0	0.2	0.4	0	0.6	−1	0	−0.6	0	−0.4	0.4	0.8	0.6	1
$C_{2.3}$	0	0	0.6	0	0	0.4	0.4	0	0.4	0	0	0	0.2	0	0	0	0.6
$C_{2.4}$	0	0.2	0.4	0	0.6	−0.6	−0.8	−0.6	0	−0.4	−0.6	0	−0.2	0	0	−0.2	0.2
$C_{2.5}$	0	0	0	0.6	0	0.4	0	0.4	0	0	0.4	0	0	0	0.2	0	0
$C_{2.6}$	0	0.6	−0.8	0.6	0.6	−0.8	0.6	0	0.6	−0.8	0	−0.6	0.2	0	0	0	0.2
$C_{3.1}$	0	0.2	0	0	0.2	−0.8	0.4	0	0	0.2	0.2	0	0.8	0.8	0.6	0.6	0
$C_{3.2}$	0	0.6	0	0	0.6	−0.6	0.4	0	0.6	0	0.4	0	0	0.6	0.8	0.8	1
$C_{3.3}$	0	0.6	0	0.2	0.4	0.4	0.4	0	0.2	0.2	0.2	0	0.6	0	0.4	0.8	0.8
$C_{3.4}$	0.8	0.8	0	0.2	0.8	−0.6	0.8	0	−0.2	0.2	0.2	0	1	0.6	0	0.6	0.4
$C_{3.5}$	0	0.4	0	0	0.6	0.4	0.8	0	0.6	0	0	0	0.6	0.6	0.6	0	0.8
$C_{3.6}$	0	0.6	0	0	0.8	0.6	1	0.8	−0.8	0.4	0.4	0	0.4	0.2	0.6	0.4	0
$C_{4.1}$	0.2	0.2	0	0.6	0.6	0.8	0.6	0.4	0.8	1	0.6	0.2	0.6	0.4	0.6	0.4	0.4
$C_{4.2}$	0.4	0.2	0.2	0.6	0.6	0.6	−0.2	0.2	0.8	0.8	1	0.6	0.4	0.4	0.4	0.4	0.4
$C_{4.3}$	0	0.8	0	0.4	0	0	0	0	−0.2	0.4	−0.2	0.6	1	0.8	0.6	0.8	0.4
$C_{4.4}$	0.2	0	0	0.2	0.2	0.4	0.2	0	0.6	0.6	0.6	0.8	0.2	0.2	0.2	0.2	0.2
$C_{4.5}$	0.2	0	0.4	0.6	0.8	−0.2	0.6	0.2	0.4	0.8	0.6	0.2	0.2	0.2	0.2	0.2	0.2
$C_{4.6}$	0	0	0.4	0.6	0.4	0.2	0.6	0.2	0.4	0.6	0.4	0.2	0.2	0.2	0.2	0.2	0.2
$C_{4.7}$	0	0	0	0.2	0	0	0.4	0	0.6	0.4	0.4	0.6	0.2	0.6	0.4	0.2	0.4
$C_{5.1}$	0	0.4	0	0	0.4	0.2	0.2	0	0.2	0.2	0.2	0	0	0.6	0.2	0	−0.4
$C_{5.2}$	0.4	0.6	0	0	0.4	0.8	0.8	0.6	0.6	0.6	0.8	0	1	0.6	0.6	1	0.6
$C_{5.3}$	0.2	0.4	0	0.4	0.4	0.8	0.8	0.6	0.8	0.6	0.8	0	0.4	0	0.2	0.8	0.6
$C_{5.4}$	0.2	0.6	0	0	0.8	0.8	0.6	0	0.6	0.6	0.6	0	0.8	0.2	0.2	0.2	0.4
$C_{6.1}$	0	0.6	0	0.4	0	0.2	0	0	0.4	0.2	0.4	0.4	0.6	0.6	0.6	0.8	0.2
$C_{6.2}$	0	0.4	0	0	0	0	0	0	−0.4	0 –	0.2	0.4	0.6	0.8	0.8	0.6	0.4
$C_{6.3}$	0	0.4	0	0	0	0	0	0	0	0	0.4	0.8	0.8	0.6	0.8	0.6	0.6
$C_{6.4}$	0	0.4	0	0.4	0	0.2	0	0.2	0.2	0	0.6	0.6	0.8	0.4	0.4	0.4	0.8
$C_{6.5}$	0.4	0.2	0	0.6	0.2	0.2	−0.2	0	0.6	0.4	0.8	0.6	0.6	0.8	1	1	1

To obtain a comprehensible simplified overview of the new system, we first reduced the graph to its kernel cut ($\alpha = 1$), this way showing only the most dominant relationships (see Fig. 4.2). In the figure, solid line arrows indicate a positive relationship (when $A \rightarrow B$ exists, an increase in factor A also results in the increase in factor B), while dashed line arrows indicate a negative relationship (i.e., a decrease in factor A also causes a decrease in factor B).

With help of the kernel section it can be clearly seen from the above graph; which ones of the subsystems of the RWMS are the most important ones from the point of view of the operation and sustainability of the system - either as positive or as negative interactions. In the following, the factors having the greatest effect on the others

Table 4.3 Refined connection matrix created by experts as a result of the workshop, Part 2

CID	$C_{4.1}$	$C_{4.2}$	$C_{4.3}$	$C_{4.4}$	$C_{4.5}$	$C_{4.6}$	$C_{4.7}$	$C_{5.1}$	$C_{5.2}$	$C_{5.3}$	$C_{5.4}$	$C_{6.1}$	$C_{6.2}$	$C_{6.3}$	$C_{6.4}$	$C_{6.5}$
$C_{1.1}$	0	0	0	0.4	0	0	−0.6	0	0.4	0.4	0.4	0	0	0	0	0.2
$C_{1.2}$	0	0.2	0.2	0	0	0.2	−0.2	−0.6	0.6	0.2	0.6	0.8	0.4	0	0.6	0.4
$C_{1.3}$	0	0.2	0	0	0.4	0.2	0	0	0	0.2	0	0.2	0	0	0	0.2
$C_{1.4}$	0.2	0.6	0.4	0.2	0.2	0	0	0.2	0.2	0.6	0	0.4	0	0.2	0	0.6
$C_{1.5}$	0	0.2	0	0	0.4	0.2	−0.2	0.6	0	0.2	0.4	0	0	0	0.2	0.8
$C_{2.1}$	0.4	0.8	0	0.2	0	0	0	0.6	0.4	0.4	0.2	0	0	0	0.2	0
$C_{2.2}$	−0.6	0.4	0	0.2	0	−0.2	0.4	0	0	1	0	0	0	0	0.4	0.4
$C_{2.3}$	−0.6	−0.6	0	0	0	0.2	0.4	0	0	0	0	0	0	0	0	0
$C_{2.4}$	0.2	−0.6	0	0.2	0.2	0.4	0.4	0	0	0	0.4	0	0	0	0	0.2
$C_{2.5}$	0.6	0.4	0	0.4	0.2	0	0	0	0	0	0	0	0	0	0	0
$C_{2.6}$	0.8	0.4	0	0.2	0	0.4	0	0.6	0.2	0.8	0.8	0	0	0	0.2	0
$C_{3.1}$	1	0.4	0	0.6	0.6	0.8	0	0	0	0.2	0	0.2	0	0.6	0	0
$C_{3.2}$	1	0	0	0.2	0	−0.4	−0.4	0	0	0	0	0	0	0	0	0
$C_{3.3}$	0	0	0	0.6	0	0	0.4	0	0	0	0	0	0	0	0	0.2
$C_{3.4}$	0.4	0	0	0.2	0	0	−0.4	0	0	0	0	0.2	0	0	0.6	0
$C_{3.5}$	0	0	0	0.2	0	0	0.4	0	0	0	0	0.2	0	0	0	0.2
$C_{3.6}$	−0.4	0	0	0.4	0	0	−0.6	0	0.6	0.6	0.6	0	0	0	0	0.8
$C_{4.1}$	0	0.8	0.4	0.6	0.8	0.8	0	0	0.8	0.4	0.4	0.6	0	0.2	0.4	0.2
$C_{4.2}$	0.8	0	0.2	0.4	0.8	0.8	0	0.6	0.8	0.6	0.4	0.4	0	0	0	0
$C_{4.3}$	0	0	0	0	0	0	0	0	0.2	0	0.6	0.6	0.8	0.2	0.8	0
$C_{4.4}$	0.4	1	0.2	0	0.6	0.6	0	0.2	0.4	0.4	0.4	0	0	0	0	0
$C_{4.5}$	0.6	0.4	0.2	0.8	0	0.6	0	0.4	0.4	0.4	0.4	0.2	0	0	0	0
$C_{4.6}$	0.8	0.4	0.2	0.8	1	0	0	0.2	0.2	0.2	0.2	0.2	0	0	0	0
$C_{4.7}$	0	0	0.4	0	0.2	0	0	0.2	0	0.2	0.2	0.4	0.4	0.6	0.4	0
$C_{5.1}$	0	0.4	0	0	0	0	0	0	0.2	0.2	0.2	0	0	0	0	0
$C_{5.2}$	0.6	1	0	0.6	0.4	0	0.2	0.8	0	0	0.6	0.8	0	0	0	0
$C_{5.3}$	0.4	0	0	0.4	0.2	0	0	0.4	1	0	0.4	0.2	0	0	0	0
$C_{5.4}$	0	0.4	0	0.4	0	0	0.2	0.8	0.8	0	0	0	0	0	0	0.4
$C_{6.1}$	0.2	0.8	0.4	0.2	0	0.6	0	−0.4	0.4	0	0.8	0	0.8	0.4	0.6	0
$C_{6.2}$	0	0	0.2	0	0	0	0	−0.2	0.4	0	0.8	1	0	0	0.4	0
$C_{6.3}$	0	0	0.2	0	0	0.4	0	−0.6	0.4	0	0.8	0.8	1	0	0.4	0
$C_{6.4}$	0	0.6	0	0	0.2	0.2	0.4	0	0	0	0.2	0	0	0.6	0	0.2
$C_{6.5}$	0.2	0.4	0	0	0.2	0.2	0.4	0.6	0.2	0.4	0.6	0	0	0.2	0.4	0

(the ones having outward effects on two to four factors) and the factors receiving the effects of other factors in the greatest extent (from two to four) are presented.

According to this the outward effect

- 'raising of professional standards' has the highest impact on all the four factors, which is followed by
- 'recovery' and 'external-internal regulatory coherence', both influencing three factors, and finally
- 'fee evolution' closes the list with the effect of two factors.

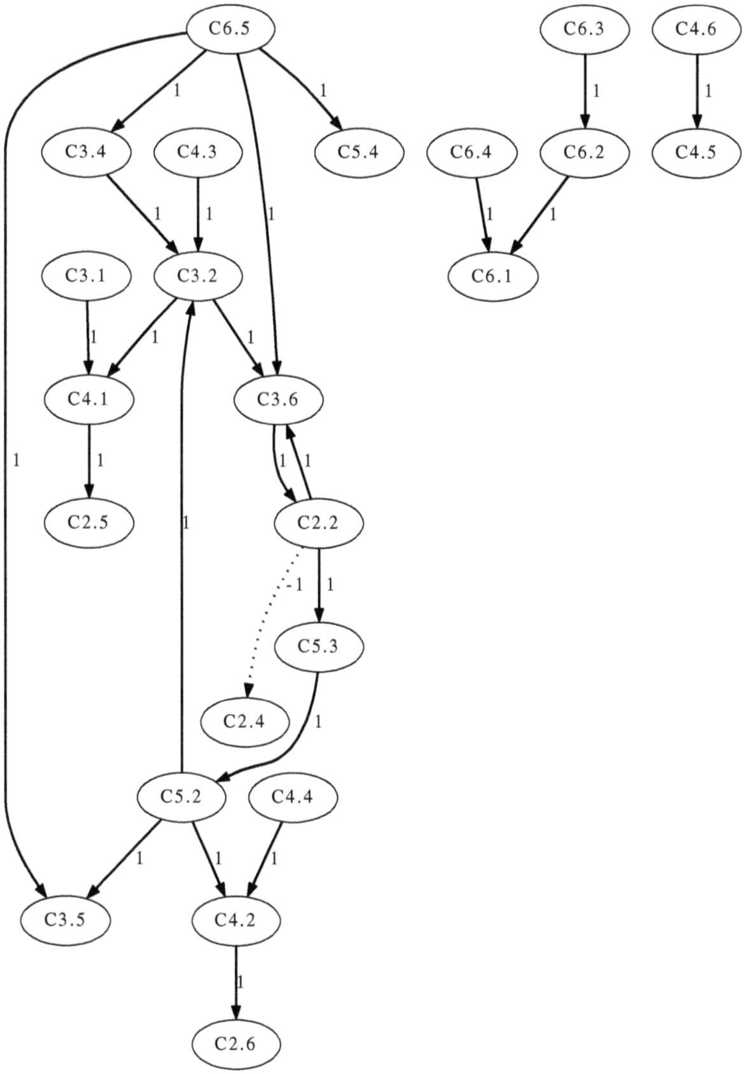

Fig. 4.2 Introduction of the most important relationships between factors and subsystems using the kernel section $\alpha = 1$

The concepts with influence received from most other subsystem elements are

- 'industrial structure' (receiving effects from three factors), then
- 'public opinion', 'health', 'funding and deductible (support)' and 'fee evolution' (receiving effects of two factors each).

It is worth noting that the 'fee evolution' subsystem element is present in both categories, i.e. according to expert opinion, it has a significant influence on the other elements, while also other system elements have dominant influence on it.

If the value of the α-cut is decreased, more and more complex graphs can be obtained, and with increasing complexity, the comprehensibility will gradually decrease. Even though the fuzzy degrees assigned to the directed edges in the graphs were discretised into a mere 11 values, it became virtually impossible to understand and evaluate the operation of such a complex system with so many interactions. The result of the workshop was 'too good', the obtained model 'too much refined', to which observation it may be added that historical statistical data are not available for the newly established concepts. In the next step of the case study, these two problems had to be dealt with, with the intention of creating a manageable technique, where the refined model could be evaluated by comparing it with some kind of data obtained from historical sources, and at the same time, the complexity of the model could be kept at a level where intuitive overview was still possible.

4.4 Summary

As a result of the workshop, based on the specialist assessment, it was possible to determine the weights and roles the interviewed experts attached to the subsystem elements (concepts) of the waste management system in terms of system operation and sustainability.

In this Chapter, a new element of the possible design procedure of modelling a waste management system was presented. Due to its complexity, this model is much more accurate in describing system components than the six-element FCM model used in the previous step of the research. The new model was created by dividing the six factors into sub-component elements. After analysing the basic model and the refined model, it emerged the fact that the two models were so conceptually different that we assumed, possibly a medium size model, between the size of the original simple and the newly obtained complex one with 33 factors could be sufficient to describe the mechanism of interactions of the RWMS in an adequate way.

Acknowledgements The paper was written with the support of the project titled 'Internationalisation, initiatives to establish a new source of researchers and graduates and development of knowledge and technological transfer as instruments of intelligent specialisations at Széchenyi István University' (project number: EFOP-3.6.1-16-2016-00017).

References

1. Applying the workshop method to effective task and problem solving, motivation and commitment of the staff. http://www.szamalk.hu/Menedzsment-es-uzleti-treningek/a-workshop-modszer-alkalmazasa-a-hatekony-feladat-es-problemamegoldas-ban-a-munkatarsak-motivaciojanak-es-elkotelezettsegenek.html
2. Cooke B, Kothari U (2001) Participation: the new tyranny? Zed Books
3. Murungweni C, Van Wijk MT, Andersson JA, Smaling EMA, Giller KE (2011) Application of fuzzy cognitive mapping in livelihood vulnerability analysis. Ecol Soc 16(4)
4. Pain Rachel (2004) Social geography: participatory research. Prog Hum Geogr 28(5):652–663
5. Wildenberg M, Bachhofer M, Adamescu M, De Blust G, Diaz-Delgadod R, Isak KGQ, Skov F, Riku V (2010) Linking thoughts to flows-fuzzy cognitive mapping as tool for integrated landscape modelling. In: Proceedings of the 2010 international conference on integrative landscape modelling and linking environmental, social and computer sciences, Montpellier, Feb 3–5. Citeseer, 2010

Chapter 5
Model Reduction Methods

Miklós F. Hatwagner

5.1 Introduction

In Chap. 3 the initial FCM model of a Sustainable Waste Management System [2] was created, which contained six concepts. These concepts were identified on the basis of consensus among the stakeholders of the field found in the relevant literature. The strength of relationships among concepts were defined by the results of a survey filled out by 75 stakeholders.

Not only the data of the model, but time series data were also collected based on the literature. It served as the input of a Bacterial Evolutionary Algorithm (BEA) to learn the connection weights among the already specified concepts and parameter λ of the threshold function. This time series data could be compared to the time series generated by simulating the dynamic behavior of the initial model. The validation of the model highlighted [4] that the time series generated by the initial model is far from the time series collected from the literature, and the model created by BEA based on time series found in literature does not resembles the initial model. Therefore, a contradiction was found between the available time series data and the initial model.

In order to resolve the experienced problem the concepts of the initial model were decomposed to further 4–7 sub-concepts according to the System-of-Systems approach as described in Chap. 4. It led to a very detailed, completely new model [3]. A workshop was organized with the help of 12 stakeholders who decided the sub-concepts and their interconnections. The result of their work is a FCM containing 33 concepts in total (Fig. 5.1). Unfortunately such an extremely complex model is often confuses experts (Fig. 5.2), and to work with them may be very laborious. Note that the number of connections is a quadratic function of the number of concepts.

Time series data was also collected for the new model as in this chapter shows it. This model is much more complex than the initial one thus the method of collecting time series data was also different: it was based on text mining. Several types of documents including EU and domestic legislation, directives, management plans and strategies was analysed covering the time interval from the 1970s to 2014.

L. T. Kóczy, *Fuzzy Cognitive Maps*, Studies in Fuzziness and Soft Computing 427,
https://doi.org/10.1007/978-3-031-37959-8_5

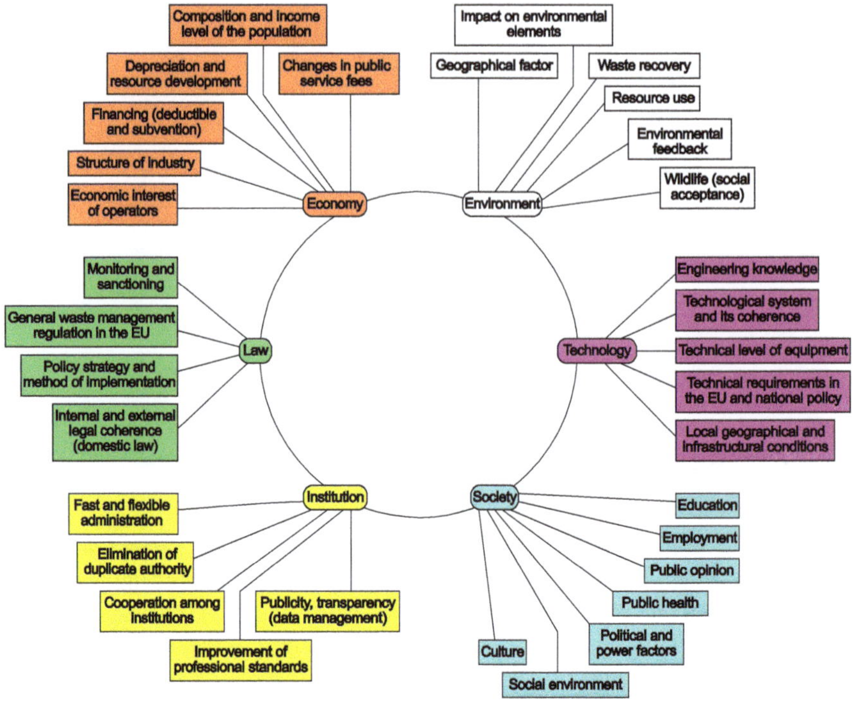

Fig. 5.1 The main concepts and their sub-concepts of regional IWMS

Fig. 5.2 The 33 concept model of regional IWMS and the relationships among its concepts. It is hard to illustrate complex FCM models appropriately and graphical model visualizations often confuse experts

The new model solved the problem of model inaccuracy and eliminated the contradiction but it is really hard or even impossible to work with such a huge model in practice. It worth noting that the new model contains 638 connections which practically cannot be handled by experts. That is why, in general, the following approach is suggested to follow in practice: start with an obviously oversized, fine-grained model. Experts are often uncertain about the importance of system components thus it worth include most or all of them in the preliminary model. Then start reducing the model automatically, in an algorithmic way until the balance of model size and required accuracy is found. The numerical accuracy of reduced models are always lower by their nature, but it does not cause a problem in practice until the decisions suggested by them are the same. On the other hand, simpler models are easier to understand, their visual representation is clearer, and it is often more important for experts and managers. In the following sections several possible ways of model reduction is presented.

5.2 Early Model Reduction Methods

Several methods had been suggested to solve the problem of oversized models before the complex model of IWMS saw the light of the day. These approaches are based on different perspectives.

In [1] an FCM is learned using historical data and its concepts are grouped into clusters in a unique way. The clustering is based on the DEMATEL [5] method. The concepts are arranged on a two dimensional plot. The vertical axis classifies concepts to cause and effect groups, the position of concepts along the horizontal axis expresses the importance of them. Based on this rearrangement of concepts two clustering methods are suggested. The first one uses K-Means clustering to create clusters according to the cause-effect behavior of concepts. The cluster centers replace the original concepts in the reduced model. The second method contains two consecutive steps, and takes also the importance of concepts into consideration. Regardless of the methods applied, experts have to define the number of clusters and they must be disjoint.

FCM was used to predict and discover knowledge about the HIV-1 drug resistance in [16]. The protease protein was modeled by FCM and causalities among sequence positions were estimated by Particle Swarm Optimization. Furthermore, Ant Colony Optimization was also applied in order to find the strongest sequence positions related to the resistance target. After that, some concepts and their connections were removed to decrease the complexity of the model until the quality of inference remained acceptable.

Another reduction method was introduced in [12]. Weak concepts and their connections were removed, then the inference capabilities of the simplified model were tested with time series data collected from real-world applications. The reduced model replaced the original if its estimation errors were acceptable.

5.3 Fuzzy Tolerance Relations-Based Reduction Methods

The first results on a novel state reduction method family were presented in [8, 10]. The members of this family differ only in the applied metric, which is used to measure the "distance" of concepts from each other. The methods may be considered as generalizations of the state reduction of finite state machines and sequential systems with partially defined states. They are widely applied in digital design where the complexity of the problem makes it possible [14]. The common idea is to create clusters of identical or similar concepts, and use these clusters instead of their members in the reduced model. The methods are based on Fuzzy Tolerance Relations (FTR), which is an extension of compatibility relations among crisp concepts. The methods examine the connection weights among concepts, because they define the effect on other concepts.

5.3.1 Description of the Main Functions Used for Reduction

At the very beginning the new model contains exactly as many clusters as the number of concepts in the original model. These clusters are all disjoint single element sets containing one of the original concepts. In the following, the ith concept is denoted by C_i, and similarly, the ith cluster is denoted by K_i. The number of concepts is n. In the next steps further concepts are added to the clusters if they are "close enough" to each member of the cluster. The "distance" of concepts can be measured by various metrics. These metrics can be selected according to the specialities of the problem, and they differentiate the members of the algorithm family. Finally, some of these expanded clusters may be identical, containing exactly the same concepts. In order to keep clusters unique, only one of these clusters is retained.

One of the main functions is called *buildCluster* (Algorithm 1). Its goal is to create a cluster that initially contains only its *initialConcept*. Later this cluster will be expanded by merger of other concepts. The second parameter, ϵ specifies the maximum allowed distance between current cluster members and the concept under examination. The value of ϵ must be in the [0, 1] interval. This parameter plays an important role in model reduction, and have to be chosen properly by experts. Low values hardly reduce the size of the model, but high values may cause oversimplification. Its appropriate value is completely problem dependent.

Function *isNearA* is called several times in *buildCluster* to decide whether the current concept C_i can become a member of cluster K or not. The number of function calls depends on how many member concepts do the cluster already have. This function (Algorithm 2) implements one of the possible metrics, but can be replaced by any of the metrics presented here. The last one was suggested in [7].

Metric "A" calculates the absolute difference of two connection weights, $w_{i,k}$ and $w_{j,k}$ (Fig. 5.3). The former one is the weight of connection between the cluster member candidate concept C_i and an independent concept C_k. The latter one is the weight of connection between C_j, which is already a member of cluster K, and C_k. Here $i \neq j \neq k$, and $i, j, k = 1, ..., n$ and n is the number of concepts. If the half of both this distance and the distance of weights in the opposite direction are below the design variable ϵ for all C_k, C_i is added to cluster K.

Algorithm 1 The *buildCluster* function

```
function BUILDCLUSTER(initialConcept, ε)
    K ← {initialConcept}
    for i ← 0; i < n; i++ do
        if i ≠ initialConcept then
            member ← true
            while member and HASNEXTELEMENT(K) do
                j ← NEXTELEMENT(K)
                member ← ISNEARA(j, i, ε)
            end while
            if member then
                K ← K + {i}
            end if
        end if
    end for
    return K
end function
```

Algorithm 2 Function *isNearA* implementing *Metric "A"*

```
function ISNEARA(i, j, ε)
    for k ← 0; k < n; k++ do
        if k ≠ i and k ≠ j then
            if |w_{i,k} − w_{j,k}| / 2 ≥ ε or |w_{k,i} − w_{k,j}| / 2 ≥ ε then
                return false
            end if
        end if
    end for
    return true
end function
```

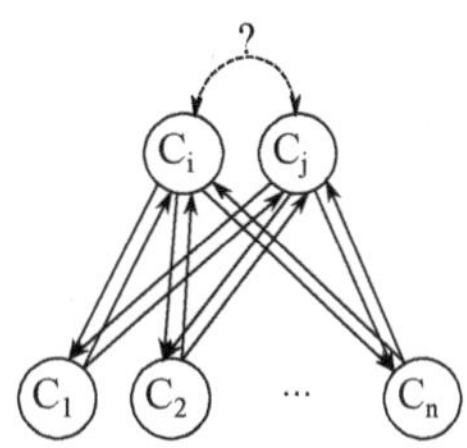

Fig. 5.3 Similarity check of two concepts, C_i and C_j [11]

Metric "B" (Algorithm 3) is a slightly modified version of Metric "A". The latter works well in most cases, but sometimes a small proportion of weight differences exceed the allowed value, and prevent the merger of concepts. The second metric provides a simple solution to this problem with the usage of parameter p. The metric allows the merger even if the distances are greater than ϵ in a small, less than p proportion of the investigated cases.

Theoretically the value of p can be any in the [0, 1] interval, but it should be rather low, however. High p values makes possible to practically merge any concepts regardless the value of ϵ, which propably leads to an unusable model.

Algorithm 3 Function *isNearB* implementing *Metric "B"*

```
function isNearB(i, j, ϵ, p)
    near ← 0
    far ← 0
    for k ← 0; k < n; k++ do
        if k ≠ i and k ≠ j then
            if |w_{i,k} − w_{j,k}| / 2 < ϵ then
                near ← near + 1
            else
                far ← far + 1
            end if
            if |w_{k,i} − w_{k,j}| / 2 < ϵ then
                near ← near + 1
            else
                far ← far + 1
            end if
        end if
    end for
    if near = 0 or far/near ≥ p then
        return false
    else
        return true
    end if
end function
```

Metric "B" allowed to lower the value of parameter ϵ with a simple trick, but the third approach (Metric "C", Algorithm 4) allows the omittance of the second parameter p by using the normalized, squared Euclidean distance. It is much easier to tune only one parameter instead of two in practice.

Algorithm 4 Function *isNearC* implementing *Metric "C"*

```
function isNearC(i, j, ϵ)
    sum ← 0
    for k ← 0; k < n; k++ do
        if k ≠ i and k ≠ j then
            sum ← sum + (w_{i,k} − w_{j,k})^2
            sum ← sum + (w_{k,i} − w_{k,j})^2
        end if
    end for
    if sum / ((n−2)*8) < ϵ then
        return true
    else
        return false
    end if
end function
```

Finally, Metric "D" (Algorithm 5) works with the Manhattan-distance of concepts.

Algorithm 5 Function *isNearD* implementing *Metric "D"*

```
function isNearD(i, j, ε)
    sum ← 0
    for k ← 0; k < n; k++ do
        if k ≠ i and k ≠ j then
            sum ← sum + |w_{i,k} − w_{j,k}|
            sum ← sum + |w_{k,i} − w_{k,j}|
        end if
    end for
    if sum / ((n−2)*4) < ε then
        return true
    else
        return false
    end if
end function
```

The applied metrics, its parameters (ϵ, p) and also the details of implementation not specified here may affect the result of reduction. For example, if the concepts provided by $nextElement$ are in various orders, the content of clusters may be different, even if the size of the reduced model remains the same. The results should be revised by experts.

The $buildCluster$ function creates one of the clusters of the reduced model. Another function, $buildAllClusters$ (Algorithm 6) calls $buildCluster$ subsequently with different $initialConcept$ values. According to the nature of the method, multiple clusters may contain the same concepts. Only one of the same clusters will be kept by the algorithm.

Algorithm 6 The *buildAllClusters* function

```
function buildAllClusters(ε)
    clusters ← {}
    for i ← 0; i < n; i++ do
        K ← buildCluster(i, ε)
        if !isElementOf(K, clusters) then
            clusters ← clusters + {K}
        end if
    end for
    return clusters
end function
```

Function $buildAllClusters$ returns the clusters, the concepts of the reduced model, but the weights among clusters have to be defined also. Function $getWeight$ (Algorithm 7) investigates the members of its parameter clusters, a and b, and calculates the average (arithmetic mean) weight of relationships between concepts included in cluster a to the concepts of cluster b. This function must be called to all possible a, b pairs to completely define the reduced model.

Algorithm 7 The *getWeight* function

```
function GETWEIGHT(a, b)
    count ← 0
    sum ← 0
    while HASNEXTELEMENT(a) do
        i = NEXTELEMENT(a)
        while HASNEXTELEMENT(b) do
            j = NEXTELEMENT(b)
            if i ≠ j then
                count ← count + 1
                sum ← sum + w_{i,j}
            end if
        end while
    end while
    if count = 0 then
        return 0
    else
        return sum/count
    end if
end function
```

Papageorgiou et al. [17] presents step-by-step the reduction process of a straightforward, five-concept model in detail.

5.3.2 Reduction of the Motivating Problem

The original model of the motivating problem (IWMS) contains 33 concepts and 638 of the theoretically possible 1056 connections, making the usage of the model very cumbersome for experts. Several experiments were conducted with different metrics and parameter (ϵ, p) values. This way the connection between reduction parameters and the evoked extent of reduction can be studied. Some interesting results are collected in Table 5.1.

Of course, the number of concepts after reduction does not characterize the usefulness of the model.

5.3.3 Theoretical Background of FTR-Based Methods

The name of the reduction methods covered in this section refer to the fact that they all based on Fuzzy Tolerance Relations (FTR). If a binary relation $R(x, x)$ is *reflexive* and *symmetric*, but *not transitive* then it is called a compatibility relation in crisp contexts and *tolerance relation* in fuzzy contexts [13]. (All further definitions in this subsection are also quoted from [13].) A fuzzy binary relation "$R(x, x)$ is *reflexive*

Table 5.1 The number of concepts in the reduced connection matrix, using different metrics [7]

Metric "A"		Metric "B"			Metric "C"		Metric "D"	
ϵ	No. of concepts	ϵ	p	No. of concepts	ϵ	No. of concepts	ϵ	No. of concepts
0.3	28	0.1	0.2	30	0.01	30	0.052	32
0.4	25	0.2	0.05	30	0.016	28	0.059	30
0.5	18	0.2	0.1	26	0.022	24	0.078	28
0.6	15	0.2	0.2	23	0.027	22	0.097	26
0.7	12	0.3	0.05	23	0.04	20	0.1	24
0.8	4	0.3	0.1	21	0.048	18	0.104	22
		0.3	0.2	15	0.054	15	0.149	20
		0.4	0.05	19	0.06	12	0.173	18
		0.4	0.1	10			0.188	15
							0.2	13

iff $R(x, x) = 1$ for all $x \in X$. A fuzzy relation is *symmetric* iff $R(x, y) = R(y, x)$ for all $x, y \in X$". The authors emphasize again that *a FTR is never transitive*. ("A fuzzy relation $R(x, x)$ is transitive (...) if $R(x, z) \geq \max_{y \in Y} \min[R(x, y), R(y, z)]$ is satisfied for each pair $\langle x, z \rangle \in X^2$.")

The four metrics defined by *isNear* functions are all real distance functions. As such, the following conditions are satisfied by function d, where $d : R(X, X) \to \mathbb{R}, \forall x, y, z \in X$:

1. $d(x, x) \geq 0$
2. $d(x, y) = 0$ iff $x = y$
3. $d(x, y) = d(y, x)$ (symmetry)
4. $d(x, z) \leq d(x, y) + d(y, z)$ (triangle inequality)

The applied metrics are symmetric functions, and *isNear* functions always return true ($R(x, x) = 1$ or $\mu = 1$) if values of the parameters are the same (reflexivity). These metrics generate non-transitive mergers, therefore they create FTRs.

Function *buildAllClusters* initiates the process of cluster building with every single concepts as *initialConcept*. The called *buildCluster* function may extend these initially single element clusters with other concepts according to the connection weights among concepts and the value of parameter ϵ. According to this behavior, the following properties hold:

- all concepts of the original model will be included in at least one cluster,
- the same concept may be included in multiple clusters, thus clusters overlap (Fig. 5.4).

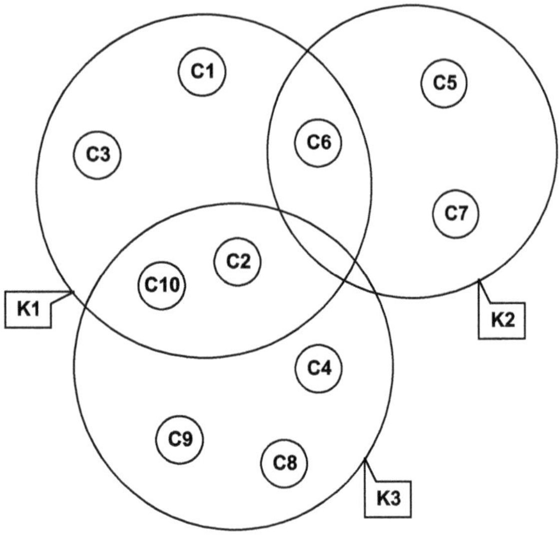

Fig. 5.4 Basic example of overlapping clusters [7]

5.3.4 *Two-Stage Learning Based Reduction*

Original and reduced models obviously behave in a different way because the latter have less concepts, the concepts represent different objects of the real system, and the relationships among concepts must be different as well. The extent of differences rely primarily on the design parameter ϵ of model reduction. Certain differences in model behaviors do not mean a problem until the same decisions can be made with both models. Unfortunately behavioral differences are hard to measure, because the values of concepts representing different objects cannot be compared directly to each other. In order to overcome this difficulty the following actions were taken in [11].

5.3.4.1 Distinction of Concept Groups

Three groups of concepts were distingiuished:

1. Concepts affecting other concepts but not affected by other concepts are called *input concepts*. They serve as the inputs of the modeled system. The states of these concepts remain the same during simulations, thus they neither need transformation function nor its parameter λ.
2. Some concepts are both affected by other concepts and they also have an effect on more or less concepts. These are the *intermediate concepts*.
3. The third group of concepts is the *output concepts* (or decision concepts). Their states are defined by other concepts, but they do not influence the state of any other concepts.

In order to make the behavioral difference of original and reduced models (the error of simulation) measurable, a modified reduction method was suggested in [11], that allows the merger of intermediate concepts only. This way simulations can be started with the same initial input concept states and the results in output concepts can be directly compared. The relation between the value of the design parameter ϵ and simulation error of the reduced model can be defined statistically using several models, many initial state vectors and ϵ values.

5.3.4.2 Modified Versions of the *getWeight* function

The *getWeight* function was also revised. The original version of it (*average*, Algorithm 7) defines the weight of connection between two clusters (a and b) as the arithmetic mean (average) weight of connections between concepts included in the corresponding clusters. More formally, the weight calculation can be expressed by Eq. 5.1.

$$w_a(a,b) = \frac{\sum_{i\in a}\sum_{j\in b} a(i,j) \times w_{ij}}{\sum_{i\in a}\sum_{j\in b} a(i,j)} \tag{5.1}$$

where $a(i,j)$ is defined as

$$a(i,j) = \begin{cases} 1 \text{ if } w_{ij} \neq 0, \\ 0 \text{ otherwise.} \end{cases} \tag{5.2}$$

The second approach (*weighted average*) calculates the weighted average of connection weights among cluster members. Greater inter-concept weights play more important role in the definition of inter-cluster weights (see Eq. 5.3).

$$w_w(a,b) = \frac{\sum_{i\in a}\sum_{j\in b} a(i,j) \times |w_{ij}| \times w_{ij}}{\sum_{i\in a}\sum_{j\in b} a(i,j) \times |w_{ij}|} \tag{5.3}$$

The third method (*extreme*, Eq. 5.4) selects the connection between clusters with the maximum absolute value, and that will be used between the clusters. If there are conncections with the same absolute value but different sign, the positive one will be chosen.

$$w_e(a,b) = \begin{cases} min(w_{ij}) \text{ if } \left|min(w_{ij})\right| > \left|max(w_{ij})\right| \\ \qquad\qquad \text{for } \forall i \in a \wedge \forall j \in b, \\ max(w_{ij}) \text{ otherwise.} \end{cases} \tag{5.4}$$

The last method (*sum*, Eq. 5.5) adds up the weight of connections among clusters. If that would not fit in the allowed $[-1, +1]$ interval, the method rounds the weight up or down to the nearest allowed value.

$$w_s(a,b) = \begin{cases} 1 & \text{if } s(a,b) > 1 \\ -1 & \text{if } s(a,b) < -1 \\ s(a,b) & \text{otherwise.} \end{cases} \tag{5.5}$$

where $s(a, b)$ is calculated as

$$s(a,b) = \sum_{i \in a} \sum_{j \in b} w_{ij} \tag{5.6}$$

5.3.4.3 λ Optimization

Model reduction inevitably causes information loss and reduced models behave less reliable. In order to somewhat compensate this problem, concepts of the reduced model did not share a common λ as in the original model but used different steepness parameters. Their values were identified by an evolutionary optimization method, the Big Bang—Big Crunch (BB-BC) [19] algorithm. The relatively low computational cost and high convergence speed made it a proper choice, moreover, the designer have to set only a few parameters to work with it. Input concepts have constant states during simulation, thus they do not need parameter λ at all. Every output concept had its own λ_{oj} parameter. It would have been the best if every intermediate concept has its own λ value, but it would have increased the computational cost of the optimization disproportionately. As a trade-off, every intermediate concept had a common λ_i parameter. According to these changes, the inference formula was modified to Eq. 5.7:

$$A_i^{(t+1)} = f_i\left(\sum_{j=1}^{M} w_{ji} A_j^{(t)} + A_i^{(t)}\right), i \neq j \tag{5.7}$$

where f_i is a sigmoidal threshold function, and its slope is specified by λ_i. These parameters were identified by the BB-BC algorithm that minimized the objective function given by Eq. 5.8:

$$\sum_{i=1}^{m} a(s_i) \times \sum_{j=1}^{n} |o_{ij} - r_{ij}| \tag{5.8}$$

Here, m denotes the number of investigated *scenarios* (initial state vectors of simulations), s_i is the ith scenario, and $a(s_i)$ is a function defined by Eq. 5.9. It expresses the fact that we are only interested in simulations leading to fixed point attractors (FPs), because only they can be taken into account in a decision making process. n is the number of output concepts, o_{ij} and r_{ij} are the values of the jth output concepts at the end of the simulation of scenario i, respectively.

$$a(s_i) = \begin{cases} 1 \text{ if } s_i \text{ resulted in fixed point attractor in} \\ \quad \text{original and reduced models,} \\ 0 \text{ otherwise.} \end{cases} \tag{5.9}$$

5.3.4.4 Test Environment and the Method of Investigation

Simulation error caused by model reduction was measured by comparing the final values of output concepts of the original and reduced models. Many synthetic models were generated to describe the effect of reduction on a statistical basis, thereafter some models found in the literature were also studied.

The original synthetic models were generated randomly, taken the following properties in consideration: in order to meet the definitions in Sect. 5.3.4.1, input concepts were forced to have at least one outgoing connection but no input connections, and similarly, output concepts had to have at least one input connection but no outgoing connections. Intermediate concepts must had at least one incoming and outgoing connections, and they were not allowed to form islands of concepts. With other words, all intermediate concepts were on at least one path leading from an input to an output concept. This property was checked by Kruskal's well-known minimum spanning tree algorithm [15]. Following Kosko's original idea, self-loops were not allowed in the generated models.

In order to imitate the diverse properties of real-life models, synthetic models with 20 and 30 concepts were generated. In the former case, there were 5 input and 3 output concepts, in the latter case 8 input and 4 output concepts were created. The density of the connection matrix may have an effect on the typical model behavior, thus models with 10, 20, 30, ..., 60 and 80% densities were generated. The meaning of *density* was defined by Eq. 5.10, however, because in our specific case some items of the connection matrix have to be always zero (all diagonal items, columns of input and rows of output concepts).

$$d = 100 \frac{n}{((c-i)(c-o) - (c-i-o))} \tag{5.10}$$

Here n stands for the number of non-zero matrix elements, c is the total number of concepts of which i are input and o are output concepts.

Simulations were performed with 25 model instances, thus finally 350 original models were involved in investigations. The common λ values of sigmoid threshold functions were also defined randomly for the test set. The simulations were started with 125 unique scenarios in case of all models.

Next, the reduction of models followed. Three ϵ values were defined for each model size–density pairs according to Table 5.2. The values were referenced by their linguistic terms (*low*, *medium* and *high*) instead of their real value, because the same

Table 5.2 Model densities and corresponding ϵ values with linguistic terms

No. of concepts	Density (%)	ϵ values		
		Low	Medium	High
20	10	0.06	0.08	0.10
20	20	0.10	0.11	0.13
20	30	0.13	0.15	0.16
20	40	0.16	0.17	0.18
20	50	0.18	0.20	0.21
20	60	0.21	0.22	0.24
20	80	0.24	0.26	0.28
30	10	0.06	0.07	0.08
30	20	0.11	0.12	0.14
30	30	0.14	0.16	0.18
30	40	0.18	0.19	0.20
30	50	0.20	0.21	0.22
30	60	0.22	0.24	0.26
30	80	0.26	0.28	0.30

parameter value causes different effect on models with different properties, and our goal was to cause comparable reduction rates.

The reduced models were created in four versions according to the various metrics applied in *getWeight* versions described in Sect. 5.3.4.2.

Due to the 2 model sizes, 7 densities, 4 weight calculation methods, 3 different reduction rates (ϵ) and 25 instances with the same properties altogether 4200 reduced model instances were obtained.

In the first stage, reduced models inherited the λ parameters of their corresponding parents. In the seconds step, however, λ-optimized models were created and not only the error caused by reduction with various ϵ parameters but the error after optimization were also determined.

If a simulation led to a FP at both the original and the corresponding reduced model, the error caused by reduction in that specific case was calculated by Eq. 5.11:

$$J_1 = \frac{\sum_{j=1}^{n} |o_j - r_j|}{n} \tag{5.11}$$

where n is the number of output concepts, o_j and r_j are the final activation values of the jth output concepts in the original and reduced models, respectively. This is a scenario-dependent error measure. In contrast, J_2 characterizes the model instance independently from its individual instances.

$$J_2 = \frac{\sum_{i=1}^{m} J_1(i)}{m} \tag{5.12}$$

Here m denotes the number of scenarios leading to FPs. The minimum, lower quartile, median, upper quartile maximum, arithmetic mean and standard deviation of J_2 errors were also calculated for reduced models with and without λ optimization. These error measures plausibly describe the information loss caused by reduction.

5.3.4.5 Results of Model Reduction

The performed tests focused on the effect of design parameter ϵ on J_2 errors. Figure 5.5 shows the results of a specific case with a box-plot: the selected models have 20 concepts, 10% densities, and the applied values of ϵ were 0.06, 0.08 and 0.1. λ optimization was not carried out. Greater ϵ values increased the median errors and the distribution of errors also, but these values remained close to zero in the investigated cases.

The next two figures (Figs. 5.6 and 5.7) gives an overview of the caused error in case of various ϵ values, weight calculation methods and model densities. As it was expected, ϵ significantly influences the caused errors, but other interesting details can also be noticed: *average* and *sum* weight calculation methods give lower errors in most cases than other methods. The advantage of using the *sum* method is especially obvious when the reduction rate is low. But if the model to be reduced is bigger and the intended rate of reduction is also greater, *average* is the suggested choice.

The mean and standard deviation of J_2 errors are easier to read in a tabulated form. Table 5.3 contains these data without considering the density of the model.

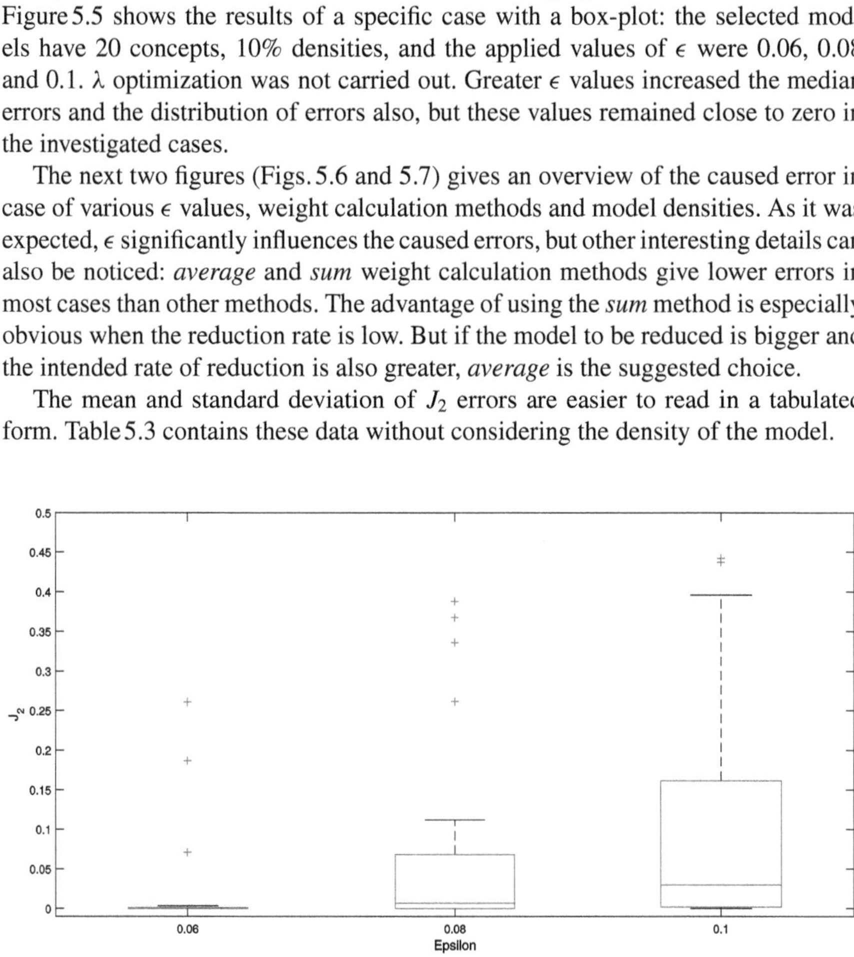

Fig. 5.5 The effect of ϵ on unsupervised reduction (without λ optimization): FCMs with 20 concepts, 10% density and after *average* weight calculation method [11]

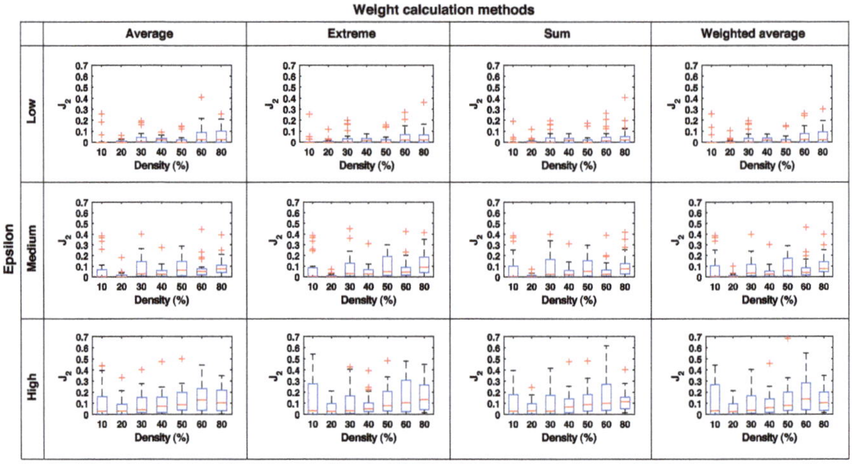

Fig. 5.6 The effect of weight calculation and ϵ in reduction for FCMs with 20 concepts [11]

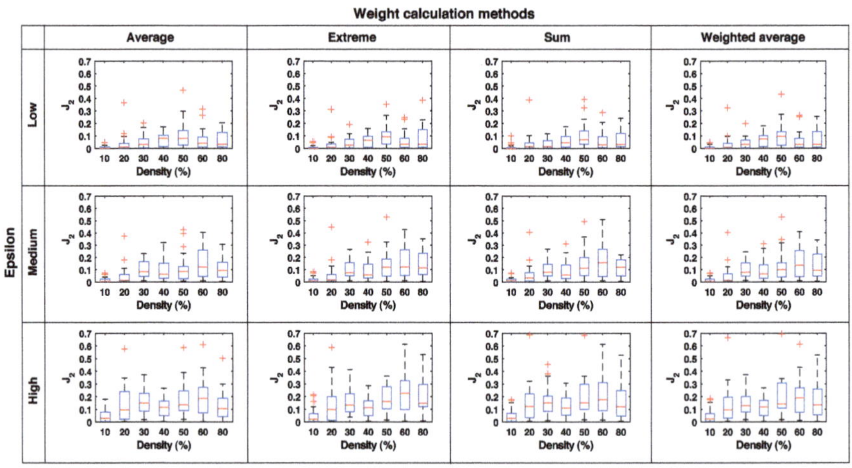

Fig. 5.7 The effect of weight calculation and ϵ in reduction for FCMs with 30 concepts [11]

Tables 5.2, 5.3 and Figs. 5.6, 5.7 give a toolset in the hands of the decision maker. For example, if he or she wants to reduce a high density model with approximately 30 concepts in a high degree, then Table 5.2 suggests to choose the value of ϵ in the [0.22, 0.3] interval. Next, according to the chosen value the weight calculation method can be selected with the help of Fig. 5.7: eg. *average* can be a reasonable choice.

If the decision maker is still discontented with the results, λ optimization can be a solution. This method is able to somewhat compensate the information loss caused

Table 5.3 Mean and standard deviation obtained after the first stage of reduction for various weight calculation methods and ϵ values [11]

Concepts	Epsilon (ϵ)	Average (10^{-3})	Extreme (10^{-3})	Sum (10^{-3})	Weighted average (10^{-3})
20	Low	46 ± 71	41 ± 62	39 ± 63	43 ± 063
	Medium	73 ± 97	81 ± 108	73 ± 100	74 ± 100
	High	110 ± 117	117 ± 129	109 ± 120	114 ± 129
30	Low	63 ± 75	59 ± 69	59 ± 72	59 ± 70
	Medium	92 ± 97	103 ± 106	101 ± 106	97 ± 104
	High	137 ± 118	157 ± 138	151 ± 135	144 ± 128

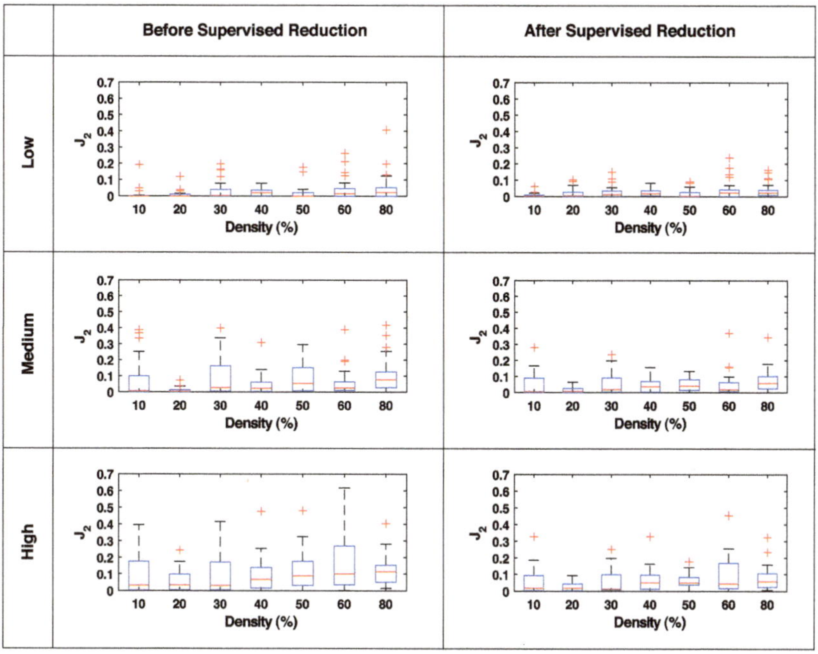

Fig. 5.8 The effect of λ optimization in reduction approach for FCMs with 20 concepts [11]

by the merger of intermediate concepts. Figures 5.8 and 5.9 makes easy to compare the errors before and after λ optimization of 20 and 30 concept models, respectively. As it can be seen, this method is especially useful when the model is highly reduced in size. The differences of mean and standard deviation of J_2 errors are numerically given in Table 5.4.

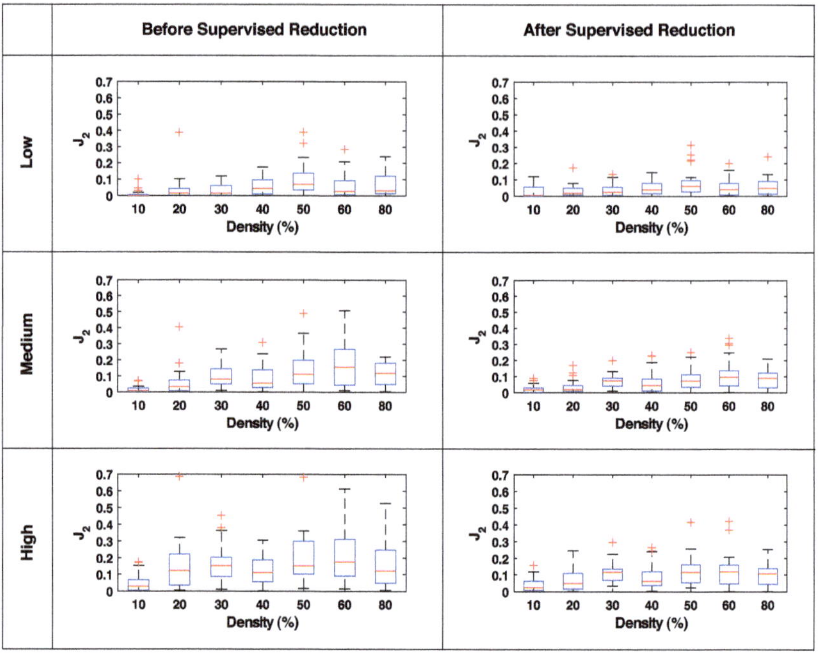

Fig. 5.9 The effect of λ optimization in reduction approach for FCMs with 30 concepts [11]

Table 5.4 Mean and standard deviation of J_2 errors obtained after λ optimization for various ϵ values and *sum* weight calculation method [11]

Number of concepts	Epsilon (ϵ)	Before λ Optimization (10^{-3})	After λ Optimization (10^{-3})
20	Low	28 ± 57	25 ± 38
	Medium	69 ± 97	48 ± 62
	High	110 ± 122	64 ± 76
30	Low	55 ± 72	50 ± 54
	Medium	101 ± 105	71 ± 67
	High	150 ± 134	97 ± 79

5.4 K-Means-Based Model Reduction Method

The well-known K-Means clustering [6] was already applied to reduce FCM models in [1], but it was used in a different way in [9]. The method, similarly to FTR-based methods, selects the concepts for merger based on their connection weights. K-Means needs data vectors for clustering, but the connection weights of an FCM are given in a matrix. Thus the first task is to create vectors from a matrix in the following way. The data vector D_i of the corresponding concept C_i is a row vector constructed from the row vector of weights representing the outgoing connections of C_i, and the transposed column vector of the weight of incoming connections. Formally, $\mathbf{D_i} = [w_{i,*}\ w_{*,i}]$, where $i \in [1, n]$ is the number of the concept, $w_{i,*}$ refers to the outgoing and $w_{*,i}$ to the incoming connection weights of C_i. This representation of connections makes possible for the K-Means algorithm to recognize the concepts with similar relationships and merge them in the same cluster.

In the examples of [9], the models followed the guidelines of Kosko and self-loops were not allowed. Consequently, all data vectors contained at least two zero-weight items, even if the method itself would have been able to handle more complex models as well.

In the case of FTR-based methods, the size of the reduced model was controlled by ϵ. In contrast, K-Means requires the number of clusters, but either way, the needed size of the reduced model can be specified. The properties of the underlying K-Means clustering causes concepts to appear only in exactly one of the clusters, and their membership value is crisp. FTR-based methods allow concepts to appear in multiple clusters thus it seems a drawback, but there are several K-Means implementations available (eg. in Matlab/Octave) making the KM-based method also easy to implement.

The connection weights among clusters were defined by Eq. 5.1.

5.5 Fuzzy C-Means-Based Model Reduction Method

Some shortcomings of the above proposed KM-based method can be compensated if Fuzzy C-Means clustering [18] is applied instead of K-Means. Similarly to FTR-based methods, it allows concepts to appear in multiple clusters, but the sum of a concept's membership values must be 1, and these values are not crisp (0 or 1).

Fuzzy C-Means also processes data vectors; these can be constructed the very same way as they were in case of the KM-based method. The size of the reduced model have to be defined in advance, too. The weight calculation method needs a slight modification, however, in order to take the various membership values into account (Eq. 5.15).

$$num = \sum_{i,j,i \neq j} \mu_{a,i} \mu_{b,j} w_{i,j} \tag{5.13}$$

$$denom = \sum_{i,j,i \neq j} \mu_{a,i} \mu_{b,j} \tag{5.14}$$

$$w(a,b) = \begin{cases} num/denom & if denom \neq 0 \\ 0 & otherwise \end{cases} \tag{5.15}$$

In Eqs. 5.13–5.15, $\mu_{a,i}$ denotes the membership value of concept C_i in cluster a, and similarly, $\mu_{b,j}$ express the membership value of C_j in cluster b. $w_{i,j}$ is the weight of connection between concepts C_i and C_j, and $w(a, b)$ is the weight of connection between the two corresponding clusters.

5.6 Comparison of the Reduced Models Created by Different Methods

First, a basic example model is reduced by KM and Fuzzy C-Means-based methods. This way, the operation of these methods is easy to follow, understand and validate the results. After that, the complex model of the motivating problem of waste management is reduced by all the three methods and some interesting properties of the reduced models are studied.

5.6.1 Reduction of the Basic Model

The model to reduce contains 4 concepts, but its reduced versions contain only 3 clusters. Figure 5.10 and Table 5.5 show the content of clusters, the connection weights among them and also the data vectors created for clustering. In Fig. 5.10c) the membership values less than 0.5 are grayed out.

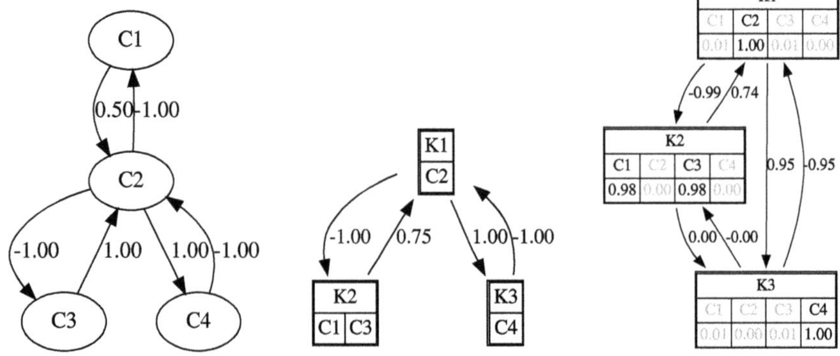

a) The original model to be reduced.

b) Model reduced with the KM-based method.

c) Model reduced with the FCM-based method. All concepts of the original model are merged into the clusters with different membership values.

Fig. 5.10 Reduction of the basic example model [9]

Table 5.5 Connection matrix and data vectors of the basic example model [9]

(a) Connection matrix of the basic example model

	C_1	C_2	C_3	C_4
C_1	0	0.5	0	0
C_2	−1	0	−1	1
C_3	0	1	0	0
C_4	0	−1	0	0

(b) D_i data vectors of concepts and their corresponding clusters

Data vector	Cluster								
D_1	K_2	0	0.5	0	0	0	−1	0	0
D_2	K_1	−1	0	−1	1	0.5	0	1	−1
D_3	K_2	0	1	0	0	0	−1	0	0
D_4	K_3	0	−1	0	0	0	1	0	0

5.6.2 *Reduction of the Motivating Model*

A basic model helps understanding the operation and behavior of a method, but in real life, the reduction of such a simple model is not necessary at all. That is why, and also to compare the reduced models created by three different methods, the reduction of the motivating problem follows. The original model contains 33 concepts. It was reduced to 18 ($\epsilon = 0.048$), 23 ($\epsilon = 0.024$) and 28 ($\epsilon = 0.018$) clusters. The details are tabulated in Tables 5.6, 5.7 and 5.8. Because of the limited space, exact membership values are not included in case of the Fuzzy C-Means-based method, concepts with greater than 0.5 membership values are indicated as cluster members instead.

The Fuzzy C-Means-based method behaved in a very interesting way. First, the concepts' membership values were always very close to their extremes (0 or 1). For example, all "high" values were at least 0.994 in the investigated cases. Second, the default parameter value of the Fuzzy C-Means algorithm ($m = 2$) resulted in models where all membership values were the same. In order to get useful models, the parameter had to be decreased to 1.1, but the algorithm is not sensitive to slightly modified values neither. The cause of this unintended effect was certainly the very long, 66-element data vectors.

The clusters of the three reduced models are rather different. Experts should be asked about the usefulness and interpretability of these models, but it would be even better to carry out a detailed behavioral analysis and comparison in the near future.

Table 5.6 Reduced models containing 18 clusters [9]

Cluster	KM-based	FCM-based	FTR-based
K1	C1.1+C6.5	C1.1+C1.3+C4.7+C6.5	C1.1+C1.2+C1.3+C2.3+C4.3+C4.4+C4.7+C5.1+C6.3
K2	C1.2+C4.3+C6.2+C6.3+C6.4	C1.2+C3.2	C1.1+C1.2+C1.4+C1.5+C3.3+C4.4+C5.2+C5.3+C5.4+C6.4+C6.5
K3	C1.3	C1.4+C1.5	C1.1+C1.2+C1.5+C3.3+C3.4+C3.5+C4.3+C5.4+C6.4
K4	C1.4	C2.1+C2.3	C1.1+C1.2+C3.2+C3.3+C3.4+C3.5+C6.1
K5	C1.5+C4.4+C4.5+C4.6	C2.2	C1.1+C1.3+C1.4+C2.5+C3.1+C3.3+C4.3+C4.4+C4.5+C4.6+C6.1
K6	C2.1	C2.4	C1.1+C1.3+C1.4+C2.5+C3.1+C3.3+C4.3+C4.4+C4.6+C6.1+C6.4
K7	C2.2	C2.5	C1.1+C1.3+C1.4+C3.1+C3.3+C4.3+C6.1+C6.2+C6.3+C6.4
K8	C2.3	C2.6	C1.1+C1.4+C1.5+C2.5+C3.3+C3.5+C4.4+C5.2+C5.3+C6.4
K9	C2.4	C3.1	C1.1+C1.4+C1.5+C4.1+C4.4+C4.5+C4.6+C5.2+C5.3
K10	C2.5+C2.6	C3.3+C3.5+C5.4	C1.1+C1.4+C2.5+C3.1+C3.3+C3.5+C4.3+C4.4+C6.1+C6.4
K11	C3.1+C3.2+C3.4	C3.4	C1.1+C1.4+C2.5+C4.2+C4.4+C4.5+C4.6+C5.2+C5.3
K12	C3.3+C3.5+C3.6+C5.4	C3.6	C1.1+C2.3+C2.5+C3.3+C3.5+C4.3+C4.4+C5.3+C6.4
K13	C4.1	C4.1+C4.4	C1.1+C3.1+C3.2+C3.3+C3.4+C3.5+C6.1
K14	C4.2	C4.2+C5.2+C5.3	C1.1+C3.1+C3.3+C3.4+C3.5+C4.3+C6.1+C6.2+C6.4
K15	C4.7	C4.3+C6.2+C6.3+C6.4	C1.2+C3.3+C3.4+C3.5+C3.6+C5.4
K16	C4.7	C4.5+C5.1	C2.1+C2.3+C2.5+C4.4+C5.1
K17	C5.2+C5.3	C4.6	C2.2+C2.3
K18	C6.1	C6.1	C2.4+C2.6

Table 5.7 Reduced models containing 23 clusters [9]

Cluster	KM-based	FCM-based	FTR-based
K1	C1.1	C1.1	C1.1+C1.4
K2	C1.2	C1.2+C6.2	C1.1+C1.5
K3	C1.3+C4.7	C1.3	C1.1+C6.5
K4	C1.4	C1.4+C1.5+C4.1+C4.4	C1.2+C6.4
K5	C1.5	C2.1+C2.3	C1.3+C4.7
K6	C2.1	C2.2	C2.1
K7	C2.2	C2.4	C2.2
K8	C2.3	C2.5	C2.3
K9	C2.4	C2.6	C2.4
K10	C2.5	C3.1	C2.5
K11	C2.6	C3.2	C2.6
K12	C3.1	C3.3+C3.4+C3.5	C3.1
K13	C3.2+C3.3+C3.5+C5.4	C3.6	C3.2
K14	C3.4	C4.2+C5.3+C5.4	C3.3+C3.4+C3.5
K15	C3.6	C4.3+C6.3	C3.6
K16	C4.1+C4.2+C4.4+C4.5+C4.6	C4.5	C4.1
K17	C4.3+C6.2+C6.3	C4.6	C4.2+C4.4
K18	C5.1	C4.7	C4.3+C6.1+C6.2+C6.3+C6.4
K19	C5.2	C5.1	C4.4+C4.5+C4.6
K20	C5.3	C5.2	C5.1
K21	C6.1	C6.1	C5.2
K22	C6.4	C6.4	C5.3
K23	C6.5	C6.5	C5.4

Table 5.8 Reduced models containing 28 clusters [9]

Cluster	KM-based	FCM-based	FTR-based	Cluster	KM-based	FCM-based	FTR-based
K1	C1.1+C5.1	C1.1	C1.1	K15	C3.4	C4.2	C3.4
K2	C1.2	C1.2+C1.5	C1.2	K16	C3.5	C4.3	C3.6
K3	C1.3+C4.7	C1.3	C1.3	K17	C3.6	C4.4	C4.1
K4	C1.4	C1.4	C1.4	K18	C4.1	C4.5	C4.2
K5	C1.5	C2.1+C2.3	C1.5	K19	C4.2	C4.6	C4.3+C6.2+C6.3+C6.4
K6	C2.1	C2.2	C2.1	K20	C4.3+C6.1	C4.7+C6.3	C4.4+C4.5
K7	C2.2	C2.4	C2.2	K21	C4.4	C5.1	C4.5+C4.6
K8	C2.3	C2.5	C2.3	K22	C4.5	C5.2	C4.7
K9	C2.4	C2.6	C2.4	K23	C4.6	C5.3	C5.1
K10	C2.5	C3.1	C2.5	K24	C5.2	C5.4	C5.2
K11	C2.6	C3.2	C2.6	K25	C5.3	C6.1	C5.3
K12	C3.1	C3.3+C3.4	C3.1	K26	C6.2	C6.2	C5.4
K13	C3.2	C3.5+C3.6	C3.2	K27	C6.3+C6.4	C6.4	C6.1
K14	C3.3+C5.4	C4.1	C3.3+C3.5	K28	C6.5	C6.5	C6.5

5.7 Summary

Complex Fuzzy Cognitive Maps are hard to use for experts for several reasons: the number of connections is the quadratic function of the number of concepts. As the number of concepts increases, it becomes very cumbersome to define the exact weights or at least the order of weights of all connections. Complex models take more time to simulate, it is harder to understand and follow the modeled processes, define the initial values of concepts, etc. On the one hand, experts are thus interested in creating small models but on the other hand, if they are not sure about the relevance of a factor they are tempted to include it in the model. This caveat can be resolved by model reduction methods: all concepts can be included in the initial model regardless of their real importance. The initial model can be reduced then by creating clusters of initial concepts until the balance of complexity and accuracy is found. These clusters can be considered the concepts of the reduced map.

Three model reduction methods were covered in this chapter, namely, the Fuzzy Tolerance Relation (FTR) based, the K-Means (KM) based, and the Fuzzy C-Means (FCM) based methods. The algorithms of methods were detailed including the possible implementations of distance functions and inter-cluster weights. Guides and tables were provided to estimate the value of design parameter ϵ to achieve the degree of reduction required. The behavioral change of models can be compensated by the optimization of λ steepness parameter of the threshold function. It was shown that by choosing a common λ for all intermediate concepts and individual λs for the output concepts the computational complexity of optimization can be maintained at an acceptable level while the behavioral changes of reduced models are minimized.

Acknowledgements The research presented in this paper was carried out as part of the EFOP-3.6.2-16-2017-00016 project in the framework of the New Széchenyi Plan. The completion of this project is funded by the European Union and co-financed by the European Social Fund.

References

1. Alizadeh S, Ghazanfari M, Fathian M (2008) Using data mining for learning and clustering FCM. Int J Comput Intell 4(2):118–125
2. Buruzs A, Pozna RC, Kóczy LT (2013) Developing fuzzy cognitive maps for modeling regional waste management systems. In: 3rd international conference on soft computing technology in civil, structural and environmental engineering, CSC 2013. Civil-Comp Press
3. Buruzs A, Hatwágner MF, Kóczy LT et al (2013) Modeling integrated sustainable waste management systems by fuzzy cognitive maps and the system of systems concept. Czasopismo Techniczne (Automatyka Zeszyt 3-AC (11) 2013):93–110
4. Buruzs A, Hatwágner MF, Pozna RC, Kóczy LT (2013) Advanced learning of fuzzy cognitive maps of waste management by bacterial algorithm. In: 2013 joint IFSA world congress and NAFIPS annual meeting (IFSA/NAFIPS), pp 890–895. IEEE
5. Gabus A, Fontela E (1973) Perceptions of the world problematique: communication procedure, communicating with those bearing collective responsibility (dematel report no 1). Technical report, Battelle Geneva Research Centre, Geneva, Switzerland

6. Hartigan JA, Wong MA (1979) Algorithm as 136: a k-means clustering algorithm. J R Stat Soc Ser C (Applied Statistics) 28(1):100–108
7. Hatwagner MF, Koczy LT (2015) Parameterization and concept optimization of FCM models. In: 2015 IEEE international conference on fuzzy systems (FUZZ-IEEE), Istanbul, Aug 2015. IEEE, pp 1–8
8. Hatwágner MF, Buruzs A, Földesi P, Kóczy LT (2014) Strategic decision support in waste management systems by state reduction in FCM models. In: International conference on neural information processing. Springer
9. Hatwágner MF, Kóczy LT (2019) Novel methods of FCM model reduction. In: Book of abstracts, p 56
10. Hatwágner MF, Buruzs A, Földesi P, Tamás Kóczy L (2015) A new state reduction approach for fuzzy cognitive map with case studies for waste management systems. In: Computational intelligence in information systems. Springer, pp 119–127
11. Hatwágner MF, Yesil E, Dodurka MF, Papageorgiou E, Urbas L, Kóczy LT (2018) Two-stage learning based fuzzy cognitive maps reduction approach. IEEE Trans Fuzzy Syst 26(5):2938–2952
12. Homenda W, Jastrzebska A, Pedrycz W (2014) Computer information systems and industrial management: 13th IFIP TC8 international conference, CISIM 2014, Ho Chi Minh City, Vietnam, 5–7 Nov 2014. Proceedings, chapter time series modeling with fuzzy cognitive maps: simplification strategies, pp 409–420. Springer, Berlin, Heidelberg
13. Klir GJ, Yuan B (1995) Fuzzy sets and fuzzy logic—theory and applications. Prentice Hall, USA
14. Kohavi Z, Jha NK (2009) Switching and finite automata theory, 3rd edn. Cambridge University Press
15. Kruskal JB (1956) On the shortest spanning subtree of a graph and the traveling salesman problem. Proc Am Math Soc 7(1):48–50
16. Nápoles G, Grau I, Bello R, Grau R (2014) Two-steps learning of fuzzy cognitive maps for prediction and knowledge discovery on the hiv-1 drug resistance. Exp Syst Appl 41(3):821–830. Methods and Applications of Artificial and Computational Intelligence
17. Papageorgiou EI, Hatwágner MF, Buruzs A, Kóczy LT (2017) A concept reduction approach for fuzzy cognitive map models in decision making and management. Neurocomputing 232:16–33
18. Yen J, Langari R (1999) Fuzzy logic: intelligence, control, and information. Prentice Hall
19. Yesil E, Urbas L (2010) Big bang-big crunch learning method for fuzzy cognitive maps. World Acad Sci Engi Technol 71:816–825

Chapter 6
Text Mining Based on the Literature of Waste Management as the Source of a Reconstructed Time Series

Adrienn Buruzs

6.1 Introduction

The main handicap of further analysis of the system under investigation was the total lack on information on the statistical behaviour of the concepts in the new, more detailed model. As a novel idea, text mining seemed to offer a source to identify the statistical behaviour needed for further research. The roots of text mining go back to the processing of medical texts, one of the first applications of text mining was in medicine [12]. Text mining is a method for analysing the contents of written documents. According to [15], the characteristics of text mining can be summarized as follows: the task of text mining is to process and interpret textual information, which makes up a large part of the data set accumulated in organizations, and to uncover new information hidden in texts. Dynamically growing textual data assets make it essential to use text mining applications. There is a myriad of powerful tools available to users, but multidisciplinary expertise is required to operate the tools.

Various text mining technologies provide multiple benefits to users: they can process huge amounts of text relatively quickly, objectively, and routine tasks can be automated (see below, in more detail). Current commercial text mining applications are complex and intelligent systems with multidisciplinary background: they use methods of information retrieval, computer language processing, data mining, and statistics, among others.

The first step in text mining is to select the text attributes to be analysed. Feature selection is one of the most critical points here, when free form and unstructured texts are transformed into a structured data set that can be handled by mathematical and computational tools. The main assumption of the analysis is based on the abstraction that the meaning contents of texts can be described by the words in them (which is obviously a simplification of the complicated semantics of general natural language texts). The method does not use all the words in the text - it omits those that are too common, have no deeper meaning, or those which are or just too rare to be used in statistics. It does not take into consideration the different grammatical forms of a word (e.g., singular or plural) separately. Word matching is a difficult but more or

L. T. Kóczy, *Fuzzy Cognitive Maps*, Studies in Fuzziness and Soft Computing 427,
https://doi.org/10.1007/978-3-031-37959-8_6

less manageable task in the case of the Hungarian language, as the latter has a rather complex morphology with plenty of suffixes and a large number of root changes, etc. When searching for text attributes, a kind of noise filtering can also be introduced in order to merge words with similar meanings into one feature. (Hungarian is one of the world's languages with the largest vocabulary and plenty of synonyms.)

Summarising the introductory thoughts, text mining is the science of extracting knowledge from unstructured or only slightly structured text files; the collection and representation of textual knowledge and information by machine intelligence methods from various document sources that remain hidden and undiscovered before processing by the analyst. Text mining is meant to offer much more than the simple search of words and the construction of statistics. While in the case of text search we want to find the contained information with as small as possible investment of time (by highly relevant search results), in the case of text mining, knowledge which was not explicitly included in the available document file (corpus) may be also gained, discovering indirect, hidden, latent information [3].

The purpose of intelligent text mining is to detect non-trivial information and to extract added information from textual data stored in textual form - on the Internet, text databases, corporate intranets or personal computers—i.e. typically unstructured. As an increasing proportion of data stored in electronic form is made up of text documents (85% of business information is unstructured data: e-mails, reminders, business and research reports, presentations, news, promotional materials, websites, etc.), there is a need for solutions that can efficiently process and analyse such texts in an intelligent way [13].

One of the primary applications of text mining is to provide a basis for decision support by gathering, condensing facts discovered in the texts, and exploring the connections of these facts. As mentioned before, the key benefits of text mining technologies are

- the ability to process huge amounts of textual data quickly, which would be an impossible task for the reader, item the objectiveness, as the results do not depend on the content of the texts, only on the applied text mining algorithms, while
- the routine tasks can be automated, giving more resources to tasks that require demanding thinking [15].

On the other hand, with the help of time series, the temporal development and regularities of socio-economic phenomena, i.e. the tendency of change and development, can be examined. Time series may be based on the nature of the data

1. expressed in the form of derived numbers obtained from the
2. absolute numbers directly measured or collected,

and according to the directions of development over time various tendencies may be recognised, such as

1. increasing,
2. decreasing,
3. linear,

4. stagnant,
5. progressive (accelerating) and
6. degressive (slowing down).

tendencies, just to mention the simplest ones. The comparability of various time series data and the requirements for the time series to be analysable are mentioned in the literature as follows:

1. the time series should cover a long enough period,
2. the data cover the same content over this period and
3. actual, real data should be used (estimations should be avoided) [8].

Time series have been used in many different areas, among other, meteorological, water level, agricultural, production, etc. data, but they play an important role in the fields of psychotherapy, education, and psychology as research tools, too, i.e., in all those disciplines where the data contain time-dependent variables. New data analysis techniques have made it possible to develop and apply a large variety of methodological approaches [9].

The essence of time series analysis is formulated by Závoti [16] as follows. Events move in a (more or less) predetermined way, where the most important task is to get to know this ordered trajectory so that the future can be predicted. Thus, in order to forecast the future, we need to know the parts and elements of the road leading there. As the time series analysed must always consist of real data in which each element in the series represents a value recorded at a given moment, thus using some predictive model, by observing certain variables in the past, it is assumed that their future values can be extrapolated [11].

After this short introduction, we return to the original problem, how to obtain time series on the concepts of the new FCM to be created, the concepts that have been identified by the team of experts, while no actual data had been recorded for those. This contradiction could be resolved by the text mining approach that was called to help in generating the missing data as objectively as it was possible at all.

6.2 The Methodological Approach

As it was mentioned earlier, it is possible to use two types of input data for the model used in FCM simulation; the simultaneous use of both types of data sets is although not essential, but we used both in this study, as the two are complementary and the results obtained from two different sources offer interesting possibilities for comparison and validation.

The two types of data sources are very different. The first input data group - the relationship matrix obtained from expert estimations - and its process of generation were presented in the previous chapters.

The other set of information is based on observations and registration, which can be considered a relatively objective technique, much more so than the previous,

Table 6.1 Time series generated from literature data

	t_0	t_1	t_2	t_3	t_4
Technology	0.20	0.35	0.60	0.75	0.80
Environment	0.15	0.20	0.40	0.60	0.80
Economy	0.10	0.15	0.30	0.50	0.70
Social	0.10	0.15	0.20	0.40	0.60
Legal	0.10	0.30	0.50	0.70	0.80
Institutional	0.10	0.20	0.30	0.50	0.60

expert knowledge based one. The trend of the examined factors was analysed on the basis of literature data in the period first 1980–2010 and then 1970–2010, and the time series was established on the basis of these, in two stages. Thus, by analysing the historical data of the factors, the team performed a retrospective reconstruction of their weights and roles.

In the first phase of the research, the data set collected for the production of the time series was considered objective, even if it was not the result of exact measurements, but based on the observations and evaluations published in the literature by researchers who were familiar with the process. It should be clearly seen that the time series-based simulation relied on objective information and no expert consultation was required. The time series for the six elements produced in the first step was created for the period 1980–2014, recorded in decades, and is repeated here, for later comparison, in Table 6.1.

Given the nature of the problem identified when validation of the time series based FCM model did not happen successfully, when it was compared with the expert knowledge based relational matrix, it was clearly necessary to modify, to refine the original model. It was recognised as an obvious fact that the simple, six-concept model so widely accepted in the literature, and considered as a consensus based approach, did not work at all. At this point, we had to recognise that a much more refined and detailed model might be the way out of this contradictory situation. So, the team decided to combine intelligent computational methods with some human knowledge and professional judgment, which is considered in the literature as the standard method of natural language processing in cases where it is difficult to determine the exact, regular "machine path" of the problem [4].

The aim of the new approach of the team was to determine the extent to which each newly identified subsystem contributed to the political decision-making process and to the operation of waste management systems. In this chapter, we present the novel approach—creating an as objective as possible time series from the results obtained by a text mining tool. We appreciate that such time series cannot be considered entirely objective as they do not rely on direct observation and registration, but they are still the most objective source available for estimating the objective values of those variables in the time series whose values have never been recorded in fact in the period in question.

As the subsystems were identified with the help of experts, with their accumulated professional experience and knowledge, it could be assumed that the theoretical model for the time series to be reconstructed would represent real and useful information. Text mining offered the most reliable tool for the reconstruction, as from the linguistic point of view, these subsystems could be considered as high-level concepts expressed by certain words and groups of words, appearing in the texts representing databases created by specialists of the field.

Each high-level concept may be associated with a collection of related words and synonyms. The knowledge base seeks to capture the relationships between concepts in a particular field, and thus represent the professional knowledge of experts by listing the relevant language units and the statements associated with them. The this way constructed knowledge bases primarily contain descriptions of relationships between concepts and words in the texts.

Although the structure of knowledge bases can typically be described as explained above, in the present research, we were not interested in the relationships between words and high-level concepts. We wanted to rely on the characteristics of the refined conceptual structure, in this way we tried to identify the contribution of each subsystem to the operation of the system based on the frequency of occurrence of the given word or phrase in the texts. Thus, the systematic relationships between the elements of the high-level concepts were omitted and the terms were simply listed under the relevant subsystems. Thus, in this phase, the task was to produce a time series for all 33 factors (new refined concepts), the earliest texts examined approximately going back to four decades. This time series shows the weight of the subsystem elements in the waste management system in each period.

Producing such time series, either based on only explicit professional literature information, or relying on the available expert judgment, would have been a cumbersome process and would have led to rather unreliable and very little objective results. On the other hand, the method of text mining promised to be suitable for the production of the time series, which allowed to extract knowledge from the unstructured or only slightly structured text files related to the topic on hand that could be found in abundancy, to gather and represent textual knowledge and information from the various document sources with machine intelligence based mining.

Our basic assumption was that this tool could be used to generate historical time series for all identified new refined concepts of the waste management system. The text mining process [6, 7, 10] in this research included the following steps:

1. definition of document selection criteria;
2. collection of documents;
3. analysis of key terms; document encoding; control; and
4. analysis.

When selecting the documents, it was necessary to consider what type and how many types of materials were included in the study. The date of entry into force of legal documents, or the issuing organisation has been taken into account in order to keep track of changes in the practice and the legislation. The time horizon of the study was defined as the period from 1970 to 2014, spanning a good 40 years. The documents themselves were mostly collected from the internet. When we could not find the material we were looking for, the university library and expert colleagues proved to be a suitable source.

The range of documents included in the analysis contained EU legislation, (framework) directives related to waste management, domestic legislation at various levels, waste management plans and strategies.

During the study, we examined the texts of nearly 115 directives, legislation and strategy documents. For each of the possible documents, we examined the extent to which it could be related to the sustainability study of the researched topic. The documents were stored in groups according to the decades of issue.

We first reviewed the range of documents collected and verified by an environmental law expert to find out what set of documents it was, what file formats we were dealing with. After this step, we were able to estimate how long it would take to convert the texts stored in the documents into a manageable format for the study.

We then compiled a list of search terms. For each of the 33 subsystem elements, we defined the range of words and expressions on the basis of which the text mining search could be started. Next, some examples are given.

1. $C_{1.2}$ Technological systems and their coherence:
    - System, system components
    - Optimal technology
    - Optimized collection and delivery
    - Volume reduction, compression
    - Waste management system
    - Selective collection
    - Deposit
    - Waste management, waste management technology
    - Type vessel, saturation of vessels
    - Frequency of emptying
    - Life cycle thinking
    - Waste hierarchy
    - Waste flow
    - Reuptake
2. $C_{2.2}$ Recovery:
    - Material utilization, energy utilization, closed cycle
    - Circulation
    - Second raw material
    - Secondary fuel
    - Energy and material recovery
    - Recycling
3. $C_{3.5}$ Financing (deductible and grant)
    - Investment financing
    - Deductible

- Support
- Application opportunities
- Government subsidies
- Dependence on subsidies

First, we used the keywords of the compiled list and measured their frequencies for all subsystems. Legal texts represent a peculiar usage of language [5], so we achieved very poor results with the method presented above; we did not find the search words or phrases at all, or only with very low frequencies. This analysis did not lead to a result as the list we compiled did not prove to be detailed enough. The negative result showed that legal texts mostly use a slightly different vocabulary as well as concepts that represent lower conceptual hierarchies of the expert knowledge base, thus a somewhat different approach had to be applied.

In order to eliminate the above problem, we enriched the search words with the help of the Hungarian WordNet® data base. For each keyword, we queried the so called synsets, grouped by relations. WordNet® is a large lexical database; nouns, verbs, adjectives, and adverbs of cognitive synonyms are grouped, each meaning a separate concept. These synsets are connected through semantic and lexical relationships. With the help of the resulting network of meaningfully related words and concepts, the texts can be analysed. WordNet® is a useful tool for computational linguistics and natural language processing. It should be mentioned that WordNet® is free to download and available to every interested user.

We further expanded the list with collocations, we retrieved two- and three-word collocations, respectively. This gave us an overview of the relationships that came with it and allowed us to determine if the particular collocation, etc. should have had a place on the list. We compared the decades with the "Relatively New Terms" method (1970/1980, 1980/1990, 1990/2000, 2000/2010). With the help of this, we learned which terms occur more significantly more often with the change of years. Using this method, we collected significant differences between the decades of bigrams (two-word sequences that often occur together) and trigrams (similar three-word sequences).

Words belonging to predefined concepts were marked and coded. After analysing the words in the text, the words were classified into the predefined concepts based on their meaning and relevance. To ensure consistency and reliability of the coding and analysis process, each word was checked by a second expert person and, if necessary, the list was modified [14]. The data thus obtained were then analysed and the trends determined.

In the next phase, the raw texts were extracted from the document files using the open source Apache Tika library [1]. The extracted text was subjected to standard pre-processing, which meant:

- all non-letter characters (e.g. punctuation) have been removed,
- all other characters have been converted to their lowercase equivalents,
- stop words were filtered out using the list of Hungarian stop words [2] (stop words are usually short conjunctions, adjectives, nouns, or strings that occur frequently

Table 6.2 Some examples of word frequency in 2010 documents

Word	Occurrence	Frequency
Municipal	247	0.070
Interest	702	0.200
Responsibiliy	141	0.040
Economic	467	0.133
Cost	326	0.093
Support	472	0.134
Producer	570	0.162
Environment	458	0.130
Consumption	455	0.129
Utilization	1020	0.290

on the Internet, such as http or com, more examples: then, enough, against, but, since, and, between, then, for, so),

- each word has undergone an "origin test", which is an automatic process to determine the root of the word; for this we used the magyarlanc2.0 package [17].

Following standard corpus linguistic methods, we determined the frequency of words in the available documents over a decade, based on "pre-processed" simple text files.

The word frequencies distributed over the decades were then assigned to the 33 subsystem elements (refined concepts) within each decade period. Thus, frequency tables for the examined decades and subsystem elements were produced, in which the role of each word and expression was determined by their frequencies of occurrence. The frequency of occurrence is obtained by dividing the number of occurrences by the total number of words in the text. Some examples of this are shown in Table 6.2.

6.3 Application of the Time Series Created to the Refined System Components Obtained from Text Mining

Based on the methodology described above, the goal at this stage of the research was to produce a time series for each of the 33 factors of the new FCM. Thus, the time series obtained by text mining showed the weights of the subsystem elements in the waste management system in each period.

An example of the time series generated for each factor and subsystem element is illustrated in Table 6.3.

The time series of the above and the other elements were used as one of the groups of input data for the systematic modelling of the waste management system.

By the method described briefly in this Chapter, a new and refined FCM model was obtained. However, a new difficulty of validating this model arose, namely, the

Table 6.3 The time series of the "Environment" factor group

	1970'	1980'	1990'	2000'	2010'
Impact on environmental elements	1.83	1.79	2.37	1.35	1.95
Utilization	1.90	0.30	0.31	0.77	1.06
Geographical factor	0.09	0.29	0.07	0.02	0.00
Resource use	0.21	0.12	0.08	0.22	0.89
Wildlife (social acceptance)	0.12	0.15	0.55	0.43	0.20
Environmental feedback	0.37	0.88	0.36	0.54	0.32

number of connections among the new concepts exceeded the number any expert could overview and evaluate. We had now a very informative model with no ways to be compared with the expert knowledge and expectations. This situation led to the next phase of the research, this time leading to a much more general direction; how to reduce in a meaningful way too complex and too large FCMs where the number of connections is causing a combinatorial explosion.

6.4 Summary

The next step in the research was to establish new time series corresponding to the new refined concepts proposed by the expert team. We found and examined the text of about 115 directives, legislation and strategy documents. Subsequently, after the classification of the word frequencies divided into decades into the 33 subsystem elements within each period, the frequency tables for the examined decades and subsystem elements were obtained, in which the role of each word and expression was determined by their frequencies. The obtained time series and the corresponding FCM model raised new and challenging questions.

Acknowledgements The paper was written with the support of the project titled 'Internationalisation, initiatives to establish a new source of researchers and graduates and development of knowledge and technological transfer as instruments of intelligent specialisations at Széchenyi István University' (project number: EFOP-3.6.1-16-2016-00017).

References

1. Apache tika. apache software foundation. https://tika.apache.org/
2. List of hungarian stop-words. http://www.tutorial.hu/hungarian-stop-words/
3. Text-mining. http://hu.wikipedia.org/wiki/Sz%C3%B5vegb%C3%A1ny%C3%A1szat
4. Allemang D, Hendler J (2011) Semantic web for the working ontologist: effective modeling in RDFS and OWL. Elsevier
5. Bhatia VK, Langton NM, Lung J, Opportunities and threats for corpus linguistics (2004) Legal discourse. Discourse in the professions: perspectives from corpus linguistics 16:203–231
6. Carley K (1990) Content analysis
7. Elo S, Kyngäs H (2008) The qualitative content analysis process. J Adv Nurs 62(1):107–115
8. Ertsey I (na) Analysis of time series. Statistics I. Lecture notes Nr 13. University of Debrecen
9. Gottman JM, McFall RM, Barnett JT (1969) Design and analysis of research using time series. Psychol Bull 72(4):299–305
10. Krippendorff K (2018) Content analysis: an introduction to its methodology. Sage Publications
11. Wei L, Yang J, Liu X (2014) Numerical prediction of time series based on FCMS with information granules. Int J Comput Commun Control 9(3):313–324
12. Swanson DR (1986) Fish oil, raynaud's syndrome, and undiscovered public knowledge. Perspect Biol Med 30(1):7–18, 1986
13. Tikk D (2005) Text-mining. Typotex
14. Wach E, Ward R (2013) Learning about qualitative document analysis
15. Wittek P (2005) Practice and difficulties of text mining. Association of Hungarian Information Brokers (2005)
16. Závoti J (2010) Mathematical statistical analyses 7. Introduction to time series analysis. University of West Hungary, TAMOP 4.2.5 Books
17. Zsibrita J, Vincze V, Farkas R (2013) Magyarlanc: a tool for morphological and dependency parsing of hungarian. In: Proceedings of the international conference recent advances in natural language processing RANLP 2013, pp 763–771

Chapter 7
Behavioral Analysis of Fuzzy Cognitive Maps

Miklós F. Hatwagner

7.1 Introduction

A few years ago, a bank visited our research team to analyze the FCMs they made. The models were related to the organizational structure and operation of their institution, but they did not disclose further details because of trade secrets. Thus, it was impossible to know exactly which processes of which organizational unit were modeled, what the meaning of each concept and edge was. Therefore we were not able to verify the correctness of the modeling process and the model itself, but we could perform various analyzes with the models. First, the static properties of models were summarized (e.g. connection matrix density, importance of concepts, cycles were detected in the connection matrix) and then different scenarios were examined, i.e. simulations were started from different initial states, and it was observed whether in these cases the system converges to a stable state (fixed point, FP), a limit cycle (LC) or a chaotic behavior (CB) occurs. (Some FPs should be achieved, others avoid according to the needs of stakeholders. LCs and CB are usually undesirable in decision making tasks.) We also examined whether the different scenarios always lead to the same fixed point. If not, we examined the number of fixed points and which cases lead to each fixed point, and in what proportion.

After answering these questions, another question arose: how reliable are the results of our analysis? If the expert who created the models e.g. neglected otherwise significant factors (not representing them with concepts) or inaccurately estimated the strength of the relationships between concepts, that may affect the outcome of simulations and lead to erroneous decisions later on. If the model was created by a single expert the model unavoidably includes his/her beliefs and subjectivity. The danger of this is less if a group of experts worked out the model instead of a sole expert [1, 9], but the result still contains some uncertainty for sure. According to our experiences they can agree relatively quickly and easily on the number of concepts, the existence and sign of relationships, but the magnitude of those relationships are much harder to define, especially if an order of importance or strength exists among them. The number of relationships is a quadratic function of the number of concepts,

L. T. Kóczy, *Fuzzy Cognitive Maps*, Studies in Fuzziness and Soft Computing 427,
https://doi.org/10.1007/978-3-031-37959-8_7

therefore even if the modelers followed Kosko's advice and they did not include self-loops in the FCM, in case of a small model containing only 10 concepts the number of relationships can be up to 90. It will be really hard to define so many weights properly and worse is that the steepness parameter λ of the threshold function does not have any connection with real objects, but its value may strongly influence the results of simulations (number and values of FPs, LCs and CB).

Because of the inherent subjectivity and uncertainty machine learning is the preferred way of modeling in most cases. In those cases it can be applied the concepts are still defined by experts, however, and/or the nature of time series data. In our case such data was not available thus the models were developed by one or more experts.

In general, a deep analysis can also be useful if the model is created by the reduction of a bigger initial model (see Chap. 6). The reduction may cause smaller or larger behavioral changes according to the degree of reduction, eg. the number of fixed-point attractors can be different.

No matter how the model is made, it worth investigating the effect of small modifications of connection weights or steepness parameter on model behavior. The sensitive points of the model can be revealed and experts can discuss about the possibilities on making the model more stable and reliable. In an FCM the connection weights are specified by real numbers, thus the number of possible modified weights are practically infinite. With a reasonable compromise we can say that the weights to try can be limited to the number of used linguistic variables, eg. 5: -1, -0.5, 0, 0.5 and 1. In the already mentioned, imagined model containing only 10 concepts, where even 90 relationships may exist, the number of model versions may up to $5^{90} = 8,078 \times 10^{62}$. Obviously the experts are primarily interested only in the effect of slight modifications, thus they would be satisfied with the check of one less and one greater values (or keeping the current one), but it could still mean $3^{90} = 8,728 \times 10^{42}$ model versions. That can be even worse if they want to take the effect of different λ values into account and of course, one simulation is not enough to characterize the behavioral effect of a modification, many simulations (depending on the model size to cover all interesting parts of the search space) are needed and these simulations are rather time consuming on their own. There is a clear need for an automated, consistent method for such investigations, and the work on that started in [4, 5] and further developed in [6, 7]. This task is too big even for a computer: an exhaustive search cannot be performed, but an evolutionary search technique, eg. the Bacterial Evolutionary Algorithm is able to guide the directions of search for a slightly modified model that behaves significantly different.

7.2 Bacterial Evolutionary Algorithm

The thorough description of Bacterial Evolutionary Algorithm (BEA) can be found eg. in [10–13], only a short summary is given here to prove the applicability and usefulness of this algorithm. BEA is an evolutionary algorithm and as such able to solve any non-linear, highly modal, high dimensional global optimization problems.

Another advantage is that it uses only a few parameters to set by the researcher. Despite its efficiency, BEA has high computational demand. It can be speed up by parallel execution [3] or by a problem specific local search operator [8], converting BEA to Bacterial Memetic Algorithm.

BEA works with a collection of possible solutions, which is also called *population*. The algorithm continuously improves the members of the population, which are also called *bacteria*, because the method imitates the reproduction and evolution of bacteria in the nature. Two operators are used to achieve this goal: *bacterial mutation* is responsible for the exploration of the search space, and *gene transfer* exploits the genetic information stored in bacteria in order to create new, hopefully fitter, more viable bacteria. The algorithm uses these operators alternately until the termination criterion is fulfilled, eg. a specified fitness value is achieved, a predefined number of generations are generated, the convergence slowed down, etc. Usually the best bacterium of the last population is considered result.

Bacterial mutation develops all bacteria at the same time, independently from each other. First, some identical copies of a bacterium are created, they are the *clones* of the original bacterium. Then every alleles of the clones are mutated, changed in random order. If any of these changes lead to a better objective value, the old allele is replaced by the new one in the original bacterium and in all the clones. This operator is elitist by its nature because it keeps the genetic information of the old bacterium if newer bacteria are less fit.

Gene transfer starts with the sort of bacteria according to their fitness values. The bacteria with better fitness go in the *superior* set, the remaining bacteria are locked up in the *inferior* set. Next, the operator selects a bacterium randomly from the superior set, and another one from the inferior set. At least one allele of the better bacterium is transferred to the worse bacterium. Then the modified bacterium have to be evaluated again, and if it has a better objective value, it has a chance to migrate into the superior set and pass on its genetic data to other bacteria during the consecutive gene transfers.

7.3 Details of the Behavior Analysis

The goal of behavior analysis is to modify a model only slightly but causing significant impact on its behavior at the same time. This can be an increased number of FPs, the presence of new LCs or CB. If such modifications are possible, they have to be further analysed by experts before the model is applied in decision making.

The search for small but significant modifications are controlled by BEA. In order to decrease the search space and accelerate the process, only 9 of the possible different weight values were tried in the last, most advanced version of the method [6], according to the applied linguistic variables (-1, -0.75, -0.5, ..., $+1$). In our experience, experts reliably recognize the lack of causal relationships, therefore zero weight connections were left untouched. It also helps to speed up the search process.

The bacteria of BEA encodes a possible, common λ value in the [0.1, 10] interval for all concepts, and all non-zero connection weights in a fixed order.

The behavioral properties of modified models were tested with several simulations started with different initial state vectors. Only 5 activation values were used: 0, 0.25, 0.5, 0.75 and 1. 1000 unique scenarios were generated randomly, and the same set was used to test all modified models for comparability. The program was prepared to automatically detect FPs, and count the number of different initial state vectors that resulted in the same FP. The simulations were definitely stopped after 100 iterations. In most cases, FCM converge very quickly to a FP. In the rare remaining cases the program searched for repeating sequences of state vectors to detect LCs. If it was not successful, the outcome of the simulation was considered a CB.

The state of a model was considered stable only if the state of all concepts were stable, viz. their values changed by at most 0.001 in the last five consecutive time steps. Unfortunately the stable states are not exactly the same in many cases due to rounding errors of floating point arithmetic or other issues. That is why the similar final states were clustered by K-Means algorithm [2] and the cluster centers were considered the common, stable, final states. K-Means needs to know the number of clusters in advance, however. It was automatically estimated by gap statistics [14].

Behavioral analysis is a multiobjective optimization problem. The sometimes conflicting goals were as follows:

1. maximize the number of FPs,
2. maximize the number of LCs,
3. maximize the number of CBs,
4. maximize the similarity of original and modified model versions.

The similarity of models was measured by Eq. 7.1.

$$d = \sum_{i=1}^{N} \sum_{j=1}^{N} (o_{ij} - m_{ij})^2 \tag{7.1}$$

where d is the difference of models, N is the number of concepts, o_{ij} is the weight of connection in the original, and m_{ij} is the weight of connection in the modified weight matrix. The fitness values of bacteria were defined in a Pareto-optimal manner. The Pareto-front of all bacteria constituted the members of the set of bacteria with highest fitness value, the Pareto-front of the remaining bacteria were the members of the set of second best bacteria, and so on.

7.4 Case Studies

A real life problem of a bank was chosen in [6, 7] to demonstrate the usefulness and operability of the developed program. Our client was interested in the result of different possible decisions regarding the organizational structure and operation which can be defined by performing FCM simulations.

The concepts of the model are collected and interpreted in Table 7.1, the causal relationships among concepts are given by Table 7.2.

First, the behavior of the original model was analyzed. The λ steepness parameter of the threshold function used in the inference equation was set to 5. Two FPs were found. 23.1% of the investigated cases resulted in the first, and the remaining 76.9% in the second FP. These FPs are pretty similar, only the values of concepts C6 and C8 differentiates them. C4 is an input concept and it was left out from Table 7.3.

Table 7.1 Concept IDs, names and categories of the first investigated model

Concept ID	Concept name	Category
C1	Clients	Assets
C2	Rules & regulations	
C3	New IT solutions	
C4	Funding	Money
C5	Cost reduction	
C6	Profit/loss	Financials
C7	Investments	
C8	Staff	Human resources
C9	New services	Product and process development
C10	Quality	
C11	Client development	Output measures
C12	Service development	
C13	Productivity	

Table 7.2 Connection matrix of the first example FCM model

	C1	C2	C3	C4	C5	C6	C7	C8	C9	C10	C11	C12	C13
C1	0	0	0.5	0	0	0.5	1	0.5	0	0.5	1	0.5	0
C2	1	0	0.5	1	0	0	1	1	0.5	0	1	1	0
C3	1	0.5	0	0	0	−1	0	−1	1	0	1	1	1
C4	0	0	0	0	0	0	0	0	0	0	0	0	0
C5	0	0	1	−0.5	0	0	0	−1	0	0	0	1	0
C6	0	0	0	0	−0.5	0	0	0	0	0	0	0	0
C7	0.5	0	0.5	1	0	0.5	0	0	0	−0.5	0	0	0
C8	0	0	0	0	0	−0.5	0	0	0	0.5	0	0	−0.5
C9	0	0	0	1	0	0.5	0.5	0.5	0	−0.5	0	0.5	0
C10	0.5	0	0	0	0	0	0.5	0.5	0.5	0	1	0	0
C11	0	0	0.5	0.5	0	0	0	0	0.5	0.5	0	0	1
C12	0	0	0.5	0.5	0	0	1	0	0.5	0	0.5	0	−0.5
C13	0	0	1	0	0	0.5	0	0	0	0	1	0	0

Table 7.3 Fixed-point attractors of the first example model

Concepts	C1–C3, C5, C7, C9–C13	C6	C8
FP#1	1.000	0.150	0.990
FP#2	1.000	0.855	0.922

The state represented by the second FP is much more desirable than the first, because it means that higher profit can be reached, while the number of staff can also be decreased. The result of every possible decisions can be checked by an FCM simulation.

The connection matrix was created by experts and may contain subjective elements and intrinsic uncertainty, however. It is hard to define the connection weight of all the 61 connections, even if only five weight values were used. If only some of these relations are estimated wrong, the result of a decision can be different than the predicted one and large financial loss may be suffered. Moreover, the value of the steepness parameter λ was chosen according to its widely applied value in literature (5), but it does not have any relationship with the modeled system. What if using another value a significantly different state will be predicted? Before realizing any intervention in the life of the organisation it worth checking the consequences of the same decision with a slightly modified model. Maybe a bit higher or lover λ value, a modified connection weight lead to a significantly different effect which must be further analysed and validated. If such a slightly modified model cannot be found, the developed model can be used with a higher degree of confidence.

The results of such an analysis are interesting even if the weights are well-chosen, because if the behavior of these modified models are known it can be seen how and where should the model be changed in order to eventuate a better operation, or what effects jeopardize the operation of it.

That is why the behavioral analysis of modified models followed. The model contains 13 concepts, but the density of it is only $\approx$39 %. It still meant a 62 variable optimization problem. The population of BEA consisted of 10 bacteria, 3 clones were created for each bacterium, and also 3 gene transfers were applied in the consecutive iterations. The optimization was stopped after the 10th generation. The connection matrices of the two best modified models (they are the members of the first Pareto-front, see Fig. 7.1) are given by Tables 7.4 and 7.5. The main properties of these models are collected in Table 7.6.

It is clear that with slight model modifications much more, over 40 possible outcomes can occur. It is vital to re-check the consequences of decisions which led to the same FP in the case of the original model.

As a second example, another model of the same bank is presented (see Table 7.7 and Fig. 7.2). Due to the confidential nature of the data, the meaning of concepts and their connections cannot be detailed here, but this is not relevant to the presentation of the applied method, neither. The behavior of the model was analysed using three different λ values first, without any modifications on the connection matrix. When

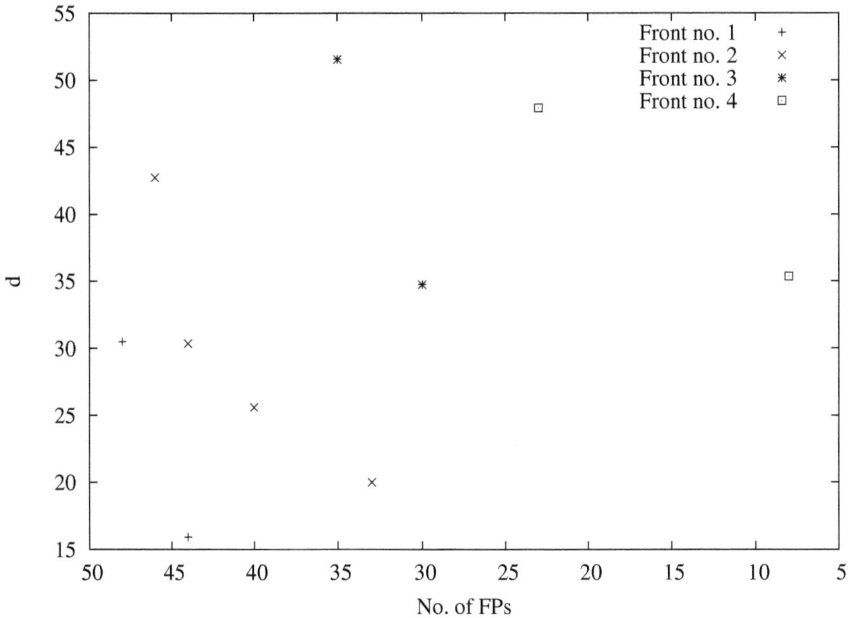

Fig. 7.1 Bacteria of Pareto-fronts in the last generation

Table 7.4 Connection matrix of the 1st model variant of the first example model

	C1	C2	C3	C4	C5	C6	C7	C8	C9	C10	C11	C12	C13
C1	0.0	0.0	0.75	0.0	0.0	0.5	1.0	0.25	0.0	0.25	0.75	0.5	0.0
C2	1.0	0.0	0.5	1.0	0.0	0.0	0.5	0.75	0.75	0.0	0.75	1.0	0.0
C3	0.25	0.0	0.0	0.0	0.0	−1.0	0.0	−0.75	0.5	0.0	0.5	0.75	1.0
C4	0.0	0.0	0.0	0.0	0.0	0.0	0.0	0.0	0.0	0.0	0.0	0.0	0.0
C5	0.0	0.0	1.0	−0.5	0.0	0.0	0.0	0.0	0.0	0.0	0.0	0.5	0.0
C6	0.0	0.0	0.0	0.0	−0.5	0.0	0.0	0.0	0.0	0.0	0.0	0.0	0.0
C7	0.75	0.0	−0.25	1.0	0.0	0.25	0.0	0.0	0.0	1.0	0.0	0.0	0.0
C8	0.0	0.0	0.0	0.0	0.0	0.25	0.0	0.0	0.0	0.75	0.0	0.0	−0.5
C9	0.0	0.0	0.0	1.0	0.0	0.25	−0.75	0.25	0.0	0.25	0.0	0.75	0.0
C10	−0.75	0.0	0.0	0.0	0.0	0.0	0.25	−0.5	0.25	0.0	−0.25	0.0	0.0
C11	0.0	0.0	0.75	0.5	0.0	0.0	0.0	0.0	0.5	0.5	0.0	0.0	1.0
C12	0.0	0.0	0.5	0.5	0.0	0.0	0.25	0.0	0.25	0.0	−0.75	0.0	−0.5
C13	0.0	0.0	0.75	0.0	0.0	0.75	0.0	0.0	0.0	0.0	0.75	0.0	0.0

Table 7.5 Connection matrix of the 2nd model variant of the first example model

	C1	C2	C3	C4	C5	C6	C7	C8	C9	C10	C11	C12	C13
C1	0.0	0.0	0.5	0.0	0.0	0.5	1.0	1.0	0.0	1.0	0.75	0.5	0.0
C2	−0.75	0.0	0.0	1.0	0.0	0.0	0.5	0.0	0.5	0.0	0.0	0.5	0.0
C3	0.25	0.25	0.0	0.0	0.0	−0.75	0.0	−1.0	0.0	0.0	0.5	−0.5	0.0
C4	0.0	0.0	0.0	0.0	0.0	0.0	0.0	0.0	0.0	0.0	0.0	0.0	0.0
C5	0.0	0.0	−0.5	−0.5	0.0	0.0	0.0	0.25	0.0	0.0	0.0	0.5	0.0
C6	0.0	0.0	0.0	0.0	0.0	0.0	0.0	0.0	0.0	0.0	0.0	0.0	0.0
C7	0.5	0.0	0.0	1.0	0.0	0.75	0.0	0.0	0.0	−0.5	0.0	0.0	0.0
C8	0.0	0.0	0.0	0.0	0.0	−1.0	0.0	0.0	0.0	1.0	0.0	0.0	0.5
C9	0.0	0.0	0.0	1.0	0.0	0.0	0.75	1.0	0.0	−0.75	0.0	0.75	0.0
C10	0.5	0.0	0.0	0.0	0.0	0.0	0.0	0.5	−0.25	0.0	−0.75	0.0	0.0
C11	0.0	0.0	−1.0	0.5	0.0	0.0	0.0	0.0	0.25	1.0	0.0	0.0	−1.0
C12	0.0	0.0	0.0	0.5	0.0	0.0	1.0	0.0	0.5	0.0	−0.25	0.0	0.25
C13	0.0	0.0	1.0	0.0	0.0	0.75	0.0	0.0	0.0	0.0	0.5	0.0	0.0

Table 7.6 Main properties of the modified model variants on Pareto front #1

Property	1st variant	2nd variant
λ value	2.366	2.070
Number of FPs	44	48
Number of LCs	0	0
Number of chaotic behavior	0	0
Difference from orig. model (d)	15.938	30.500

Table 7.7 Connection matrix of the second example FCM model

	C1.2	C2.4	C2.5	C3.1	C3.2	C4.1	C5.2	C5.3	C6.1	C6.2	C6.3
C1.2	0	0	0	0.5	1	0.5	0	0.5	1	0.5	0
C2.4	0	0	0	0	0	0	0	0	0	0	0
C2.5	0	−0.5	0	0	0	−1	0	0	0	1	0
C3.1	0	0	−0.5	0	0	0	0	0	0	0	0
C3.2	0.5	1	0	0.5	0	0	0	−0.5	0	0	0
C4.1	0	0	0	−0.5	0	0	0	0.5	0	0	−0.5
C5.2	0	1	0	0.5	0.5	0.5	0	−0.5	0	0.5	0
C5.3	0.5	0	0	0	0.5	0.5	0.5	0	1	0	0
C6.1	0	0.5	0	0	0.5	0	0.5	0.5	0	1	1
C6.2	0	0.5	0	0	1	0	0.5	0	0.5	0	−0.5
C6.3	0	0	0	0.5	0	0	0	0	1	0	0

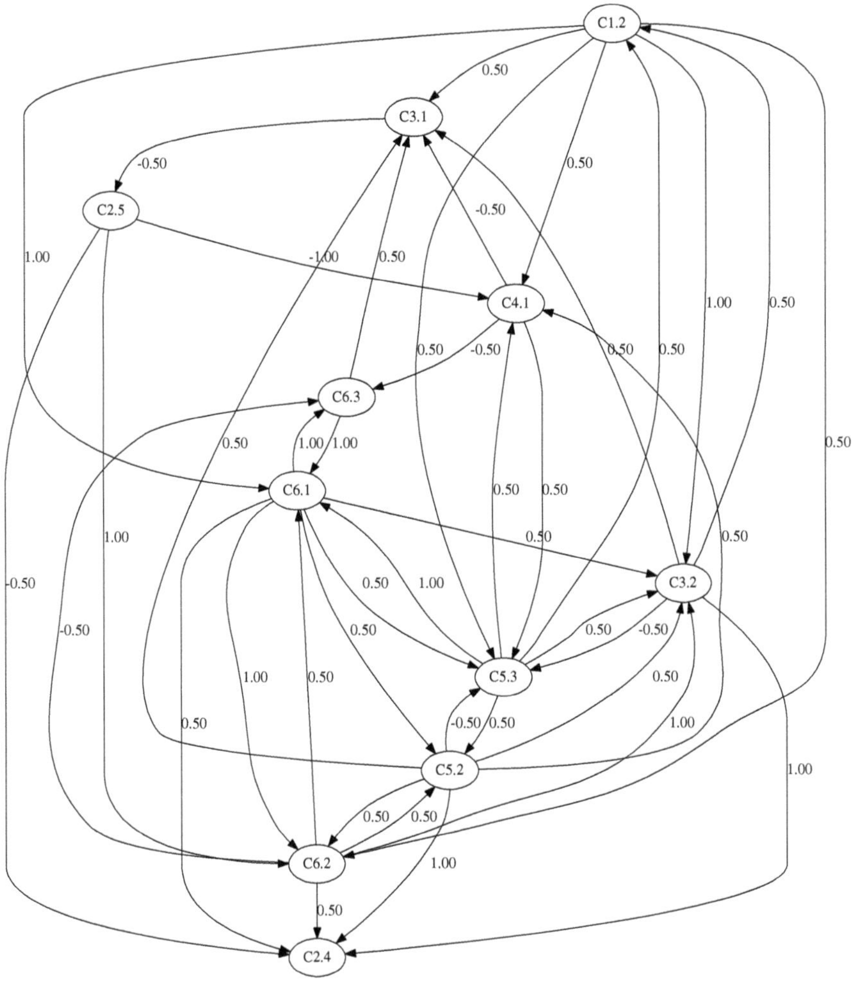

Fig. 7.2 Graph representation of the second example model

the steepness parameter was set to 5, three FPs were found (see Table 7.8). The last one occurred only in 0.2% of the investigated cases. This raises the suspicion that this is not a real fixed point, but an extremely slow convergence to another fixed point happened, which was mistakenly perceived by the software as a FP. However, repeating the study with a reduced λ value of 4.7, it can be seen that the incidence of the same fixed point increased to 8.2%, while the activation value of concept C2.5 left practically the same (0.506 vs. 0.502). Even some inital state vectors are listed in Table 7.10 as examples that lead to the specified FPs. Further reducing the value of the λ parameter to 1.4235, only one fixed point remained (see Table 7.11). These examinations show how important the value of parameter λ is in FCM simulations.

Table 7.8 Fixed-point attractors of the second example model ($\lambda = 5$)

Concepts	C1.2, C2.4	C2.5	C3.1, C3.2, C4.1, C5.2	C5.3	C6.1, C6.2	C6.3
FP#1 (25.6%)	1.000	**0.145**	1.000	0.999	1.000	0.993
FP#2 (74.2%)	1.000	**0.855**	1.000	0.999	1.000	0.993
FP#3 (0.2%)	1.000	**0.506**	1.000	0.999	1.000	0.993

Table 7.9 Fixed-point attractors of the second example model ($\lambda = 4.7$)

Concepts	C1.2, C2.4	C2.5	C3.1, C3.2, C4.1, C5.2	C5.3	C6.1, C6.2	C6.3
FP#1 (21.8%)	1.000	**0.198**	1.000	0.999	1.000	0.991
FP#2 (70.0%)	1.000	**0.814**	1.000	0.999	1.000	0.991
FP#3 (8.2%)	1.000	**0.502**	1.000	0.999	1.000	0.991

Table 7.10 Examples of initial state vectors that eventuate different fixed point attractors for the second example model ($\lambda = 4.7$)

Concepts	C1.2	C2.4	C2.5	C3.1	C3.2	C4.1	C5.2	C5.3	C6.1	C6.2	C6.3
FP#1	0.083	0.455	0.222	0.816	0.133	0.454	0.497	0.736	0.770	0.111	0.606
FP#2	0.262	0.963	0.284	0.293	0.424	0.243	0.945	0.521	0.562	0.098	0.612
FP#3	0.881	0.054	0.216	0.384	0.969	0.189	0.594	0.944	0.653	0.187	0.679

Table 7.11 Fixed-point attractor of the second example model ($\lambda = 1.4235$)

Concepts	C1.2	C2.4	C2.5	C3.1	C3.2	C4.1	C5.2	C5.3	C6.1	C6.2	C6.3
FP#1	0.934	0.995	0.510	0.965	0.998	0.929	0.968	0.866	0.997	0.993	0.755

The effect of any planned interventions must be checked by several simulations using various steepness parameters in order to realize all the possible outcomes (Table 7.9).

The next series of experiments were conducted with a bit modified connection matrices (see Tables 7.12 and 7.14) while keeping the steepness parameter value at a constant, $\lambda = 5$ level. These changes resulted in the number of fixed points being increased to 6 (see Table 7.13) and 7 (Table 7.15), respectively. It is an interesting fact that in neither case does the difference in the activation values of the concept C2.5 differentiate the fixed points from each other, as we saw in the previous series of experiments.

Table 7.12 Connection matrix of the 1st model variant of the second example model

	C1.2	C2.4	C2.5	C3.1	C3.2	C4.1	C5.2	C5.3	C6.1	C6.2	C6.3
C1.2	0	0	0	0.5	0.75	−0.25	0	−0.5	1	0.25	0
C2.4	0	0	0	0	0	0	0	0	0	0	0
C2.5	0	−0.5	0	0	0	0.25	0	0	0	−0.5	0
C3.1	0	0	0.75	0	0	0	0	0	0	0	0
C3.2	−0.75	−0.75	0	−0.25	0	0	0	−0.25	0	0	0
C4.1	0	0	0	0.5	0	0	0	−0.25	0	0	−1
C5.2	0	−0.25	0	−0.75	1	−0.5	0	−1	0	0.75	0
C5.3	0	0	0	0	0.75	−1	1	0	0.25	0	0
C6.1	0	0.5	0	0	−0.75	0	−0.5	0.25	0	−0.25	0
C6.2	0	1	0	0	1	0	0.5	0	−1	0	−1
C6.3	0	0	0	0.25	0	0	0	0	0.25	0	0

Table 7.13 Fixed-point attractors of the 1st model variant of the second example model ($\lambda = 5$)

Concepts	C1.2	C2.4	C2.5	C3.1	C3.2	C4.1	C5.2	C5.3	C6.1	C6.2	C6.3
FP#1	0.03	**0.86**	1.00	**0.86**	1.00	0.97	1.00	0.00	0.01	1.00	0.00
FP#2	0.03	**0.86**	1.00	**0.14**	1.00	0.97	1.00	0.00	0.01	1.00	0.00
FP#3	0.03	**0.15**	1.00	**0.14**	1.00	0.97	1.00	0.00	0.01	1.00	0.00
FP#4	0.03	**0.15**	1.00	**0.86**	1.00	0.97	1.00	0.00	0.01	1.00	0.00
FP#5	0.03	**0.86**	1.00	**0.50**	1.00	0.97	1.00	0.00	0.01	1.00	0.00
FP#6	0.03	**0.49**	1.00	**0.85**	1.00	0.97	1.00	0.00	0.01	1.00	0.00

Table 7.14 Connection matrix of the 2nd model variant of the second example model

	C1.2	C2.4	C2.5	C3.1	C3.2	C4.1	C5.2	C5.3	C6.1	C6.2	C6.3
C1.2	0	0	0	−0.25	0.25	0.75	0	−0.25	0	0.5	0
C2.4	0	0	0	0	0	0	0	0	0	0	0
C2.5	0	−0.75	0	0	0	−0.25	0	0	0	1	0
C3.1	0	0	1	0	0	0	0	0	0	0	0
C3.2	0.25	0.25	0	1	0	0	0	1	0	0	0
C4.1	0	0	0	−1	0	0	0	1	0	0	0.25
C5.2	0	0.5	0	0.75	−1	−0.25	0	0.25	0	0.75	0
C5.3	−1	0	0	0	1	−0.25	−1	0	1	0	0
C6.1	0	0	0	0	−0.25	0	0.5	−0.5	0	0	0
C6.2	0	−0.25	0	0	0.5	0	0	0	−0.25	0	0
C6.3	0	0	0	−0.75	0	0	0	0	0.5	0	0

Table 7.15 Fixed-point attractors of the 2nd model variant of the second example model ($\lambda = 5$)

Concepts	C1.2	C2.4	C2.5	C3.1	C3.2	C4.1	C5.2	C5.3	C6.1	C6.2	C6.3
FP#1	0.03	**0.96**	1.00	**1.00**	1.00	**0.04**	**0.86**	1.00	1.00	1.00	**0.99**
FP#2	0.03	**0.04**	1.00	**1.00**	1.00	**0.12**	**0.15**	1.00	1.00	1.00	**0.99**
FP#3	0.03	**0.04**	1.00	**0.90**	1.00	**0.82**	**0.14**	1.00	1.00	1.00	**1.00**
FP#4	0.03	**0.04**	1.00	**0.27**	1.00	**0.82**	**0.14**	1.00	1.00	1.00	**1.00**
FP#5	0.03	**0.85**	1.00	**1.00**	1.00	**0.06**	**0.50**	1.00	1.00	1.00	**0.99**
FP#6	0.03	**0.50**	1.00	**1.00**	1.00	**0.06**	**0.49**	1.00	1.00	1.00	**0.99**
FP#7	0.03	**0.15**	1.00	**1.00**	1.00	**0.06**	**0.50**	1.00	1.00	1.00	**0.99**

7.5 Summary

The developed program has reached its main goal: it finds several slightly modified models that behave differently: have more FPs, LCs, or behave chaotically. The number of FPs, the final activation values of concepts in FPs are defined, moreover, the scenarios leading to FPs, LCs, or chaotic behavior are revealed as well. These results, especially those that apply to FPs are very interesting for decision-makers because they represent the stable states of the modeled system. Some of the stable states should be avoided, others are desirable. The experts can study which scenarios lead to these states and they can make well-founded decisions that lead to one of the desired states according to the analysis. In most cases, LCs and CB are undesirable outcomes as they represent unstable, hardly predictable operations.

Even if the results are promising, there are still some problems with the method to be solved or deficiencies to be improved. One suspicious detail is the extremely high number of FPs in modified models. The applied algorithm of cluster number estimation should be reviewed because in the case of high dimensional spaces often very similar state vectors are classified in different clusters thus the number of FPs is overestimated.

The other, more serious problem is the time requirement of computations. The root cause of this is that the behavioral analysis is based on BEA, and evolutionary optimization algorithms are compute-intensive in general. In this case, the objective function to be evaluated is also very complex: the behavioral properties of a specific, modified model variant (an individual/bacterium of a population) can be discovered only by a long series of FCM simulations.

The speed of computing can be increased by the implementation of some advanced technologies, eg. parallel computing or using GPUs where they are applicable. Despite these, the method itself remains computation-heavy, and there is still a need for the quick characterization of models in an analytical way. This demand led to a thorough study of FCM models with mathematical tools.

Acknowledgements The research presented in this paper was carried out as part of the EFOP-3.6.2-16-2017-00015 project in the framework of the New Széchenyi Plan. The completion of this project is funded by the European Union and co-financed by the European Social Fund.

References

1. Groumpos PP (2010) Fuzzy cognitive maps: Basic theories and their application to complex systems. In: Fuzzy cognitive maps. Springer, pp 1–22
2. Hartigan JA, Wong MA (1979) Algorithm as 136: a k-means clustering algorithm. J R Stat Soc Ser C (Applied Statistics) 28(1):100–108
3. Hatwagner M, Horvath A (2011) Parallel gene transfer operations for the bacterial evolutionary algorithm. Acta Technica Jaurinensis 4(1):89–111
4. Hatwágner MF, Kóczy LT (2016) Uncertainty tolerance and behavioral stability analysis of fixed structure fuzzy cognitive maps. ESCIM 2016, p 15
5. Hatwágner MF, Niskanen VA, Kóczy LT (2017) Behavioral analysis of fuzzy cognitive map models by simulation. In: 2017 joint 17th world congress of international fuzzy systems association and 9th international conference on soft computing and intelligent systems (IFSA-SCIS). IEEE, pp 1–6
6. Hatwagner MF, Vastag G, Niskanen VA, Kóczy LT (2018) Improved behavioral analysis of fuzzy cognitive map models. In: International conference on artificial intelligence and soft computing. Springer, pp 630–641
7. Hatwágner MF, Vastag G, Niskanen VA, Kóczy LT (2019) Banking applications of fcm models. In: Trends in mathematics and computational intelligence. Springer, pp 61–72
8. Kóczy LT, Földesi P, Tüű-Szabó B (2018) Enhanced discrete bacterial memetic evolutionary algorithm-an efficacious metaheuristic for the traveling salesman optimization. Inf Sci 460:389–400
9. Kosko B (1988) Hidden patterns in combined and adaptive knowledge networks. Int J Approx Reason 2(4):377–393
10. Nawa NE, Furuhashi T (1998) Bacterial evolutionary algorithm for fuzzy system design. In: SMC'98 Conference Proceedings. 1998 IEEE international conference on systems, man, and cybernetics (Cat No 98CH36218), vol 3. IEEE, pp 2424–2429
11. Nawa N, Furuhashi T (1998) A study on the effect of transfer of genes for the bacterial evolutionary algorithm. In: 1998 second international conference on Knowledge-Based intelligent electronic systems, 1998. Proceedings KES'98, vol 3. IEEE, pp 585–590
12. Norberto Eiji Nawa and Takeshi Furuhashi (1999) Fuzzy system parameters discovery by bacterial evolutionary algorithm. IEEE Trans Fuzzy Syst 7(5):608–616
13. Nawa NE, Hashiyama T, Furuhashi T, Uchikawa Y (1997) A study on fuzzy rules discovery using pseudo-bacterial genetic algorithm with adaptive operator. In: IEEE international conference on evolutionary computation. IEEE, pp 589–593
14. Tibshirani R, Walther G, Hastie T (2001) Estimating the number of clusters in a data set via the gap statistic. J R Stat Soc: Ser B (Statistical Methodology) 63(2):411–423

Chapter 8
Fuzzy Cognitive Maps: Mathematical Challenges

István Á. Harmati

8.1 Introduction

In the last two decades fuzzy cognitive maps became a successful and widely-used modelling tool in describing the qualitative behaviour of complex systems [11]. The fields of applications show a large variety: risk analysis [8], medical diagnosis [1], supply chain management [12] etc. just to mention a few examples. The most advantageous features are the visualization of the causal connections between the factors of the system and flexibility in modifying some parameters and characteristics of the model [6]. On the other hand, we have to mention the mathematical knowledge required from the user: it is extremely low in contrast to other modelling methodologies (for example differential equations). Consequently, in application of fuzzy cognitive maps, the scientist can concentrate on the problem studied, instead of the mathematical difficulties of the applied methodology. Nevertheless, it is a double-edged sword: since she is not faced with mathematical difficulties in connection with the FCM method, she may think that there aren't any traps in the field. In this chapter, we introduce some mathematical questions related to FCM based modelling.

8.2 Reformulating the Updating Rule

Kosko's original updating rule [5] was the following (using the usual notation), where $A_i(k)$ is the value of concept C_i at discrete time k, w_{ij} is the weight of the connection from concept C_j to concept C_i

$$A_i(k) = f\left(\sum_{j=1, j\neq i}^{n} w_{ij} A_j(k-1)\right) \tag{8.1}$$

Here the case $j \neq i$ is excluded from the summation, but in fact it is a redundant notation, due to the fact that $w_{ii} = 0$, since early fuzzy cognitive maps did not apply self-feedback.

L. T. Kóczy, *Fuzzy Cognitive Maps*, Studies in Fuzziness and Soft Computing 427,
https://doi.org/10.1007/978-3-031-37959-8_8

Latter, the following updating rule has been introduced [2], where the current state of a concept is taken into consideration in the computation of the activation value at the next time instant:

$$A_i(k) = f\left(\sum_{j=1, j\neq i}^{n} w_{ij} A_j(k-1) + A_i(k-1)\right) \tag{8.2}$$

Note that this equation can be rewritten in a more compact form, if we introduce the following notation:

$$w_{ij}^* = \begin{cases} w_{ij} & \text{if } i \neq j \\ 1 & \text{if } i = j \end{cases} \tag{8.3}$$

i.e. we introduce $w_{ii}^* = 1$. Using this notation we have

$$\begin{aligned} A_i(k) &= f\left(\sum_{j=1, j\neq i}^{n} w_{ij} A_j(k-1) + A_i(k-1)\right) \\ &= f\left(\sum_{j=1, j\neq i}^{n} w_{ij}^* A_j(k-1) + w_{ii}^* A_i(k-1)\right) \\ &= f\left[\sum_{j=1}^{n} w_{ji}^* A_i\right], \end{aligned} \tag{8.4}$$

Let us denote the vector of activation values at time step k by $A(k)$, and denote the identity matrix by I. Then the previous equation in vector-matrix form will be

$$\begin{aligned} A(k) &= f\left(WA(k-1) + A(k-1)\right) \\ &= f\left(WA(k-1) + IA(k-1)\right) \\ &= f\left((W+I)A(k-1)\right) = f\left(W^* A(k-1)\right) \end{aligned} \tag{8.5}$$

Here the transformation function f is applied coordinate-wise.

The more general form of the updating rule deals with weighted self-feedback:

$$A_i(k) = f\left(\sum_{j=1, j\neq i}^{n} w_{ij} A_j(k-1) + d_i A_i(k-1)\right) \tag{8.6}$$

In this form, d_i $(0 \leq d_i \leq 1)$ expresses the strength of possible self-feedback. If $d_i = 0$, then we get Kosko's original updating rule, while if $d_i = 1$ we get the updating rule discussed above. Let us introduce a notation again:

$$w_{ij}^* = \begin{cases} w_{ij} & \text{if } i \neq j \\ d_i & \text{if } i = j \end{cases} \tag{8.7}$$

Then the updating rule has the form

$$A_i(k) = f\left[\sum_{j=1}^{n} w_{ji}^* A_i\right] \tag{8.8}$$

or in vector-matrix form:

$$A(k) = f\left(W^* A(k-1)\right) \tag{8.9}$$

where W^* is the sum of weight matrix W and a diagonal matrix with d_is in its diagonal i.e.

$$W^* = W + \text{diag}[d_i], \tag{8.10}$$

where diag$[d_i]$ is a diagonal matrix with d_is in its diagonal and zeros elsewhere, W is the original weight matrix, which contains zeros in the main diagonal. Note that if $d_i = 0$ for every $i = 1 \dots n$, then $W^* = W$, which yields Kosko's model. If $d_i = 1$ for every $i = 1 \dots n$, then we have $W^* = W + I$, which is the model described by Eq. 8.2. In the mathematical analysis of fuzzy cognitive maps (and its extensions) we are going to use this form of the updating rule, which means that the if self-feedback is applied, then it is incorporated into the weight matrix. Finally, the updating rule reduces to the following form:

$$A(k) = G(A(k-1)), \tag{8.11}$$

where mapping G depends on the weights (W) and transformation function(s) (f) applied. From the pure mathematical point of view, it is a so-called discrete dynamical system. There are many advanced textbooks discussing comprehensively the complex and exciting behaviour of discrete dynamical systems, see for example [15] or [7]. The complexity of the behaviour highly depends on the nature of the transformation function.

The most widely used threshold functions are the

- bivalent

$$f(x) = \begin{cases} 0 & \text{if } x \leq 0.5 \\ 1 & \text{if } x > 0.5 \end{cases} \tag{8.12}$$

- trivalent

$$f(x) = \begin{cases} -1 & \text{if } x < 0 \\ 0 & \text{if } x = 0 \\ 1 & \text{if } x > 0 \end{cases} \tag{8.13}$$

- log-sigmoid with parameter $\lambda > 0$

$$f(x) = \frac{1}{1 + e^{-\lambda x}} \tag{8.14}$$

- and hyperbolic tangent with paramater $\lambda > 0$

$$f(x) = \tanh(\lambda x) = \frac{e^{\lambda x} - e^{-\lambda x}}{e^{\lambda x} + e^{-\lambda x}} = \frac{1 - e^{-2\lambda x}}{1 + e^{-2\lambda x}} \tag{8.15}$$

In the case of bivalent and trivalent functions, there are only finite number of possible states [14]. Since the reasoning follows a deterministic updating rule, the iteration may end in a fixed point (equilibrium point) or in a limit cycle. This later one means that several states are repeated in cyclic order. In contrary, continuous threshold functions may produce much more interesting and difficult behaviour, namely [3, 9, 10, 13]:

- the iteration may lead to a fixed point;
- the iteration may arrive to a limit cycle;
- the iteration never stabilizes, which is usually called 'chaotic' in FCM literature. Nevertheless, in many cases this behaviour is not chaotic in the dynamical sense, but shows a deeply structured, more complex pattern compared to fixed points or limit cycles.

8.3 Examples for Different Types of Behaviour of FCMs

In this section, we show some examples for the different limiting behaviours of the iteration. For demonstration purposes, we choose a small network with only three nodes, to point out that complex behaviour is not necessarily matter of scale, it may occur in relatively small networks, too.

The topology of the map is shown in Fig. 8.1. As we can see, we do not apply self-feedback. The transformation function is a hyperbolic tangent function (for further interesting properties of hyperbolic tangent FCMs see [4]):

$$f(x) = \frac{1 - e^{-4.5x}}{1 + e^{-4.5x}} = \tanh(2.25\,x). \tag{8.16}$$

The initial activation vector (starting value) is $A(0) = [0, 1, 0]^T$. Weight setting is *almost* the same in the three examples, only the sign of one (or two) weights change, but it results in qualitatively different patterns.

Different settings of weights and parameter λ yield completely different behaviour, although the topology remains the same. For a certain set of parameters we may have a non-trivial fixed point (the trivial fixed point is the zero vector, since it is always a fixed point of hyperbolic tangent FCMs, but not always attractor [4]) (Fig. 8.5). Other setting yields oscillation, namely a quasiperiodic behaviour (Fig. 8.4).

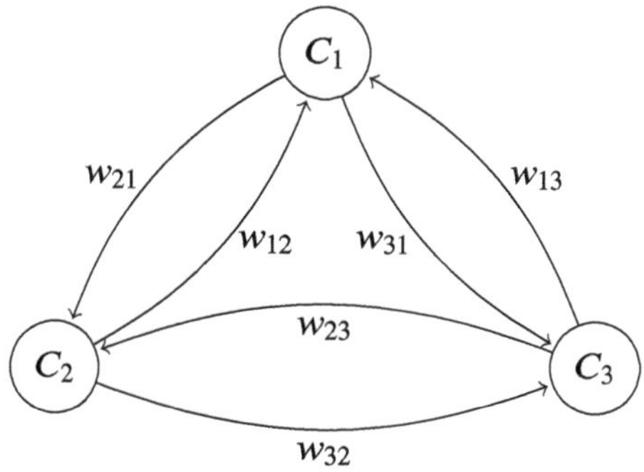

Fig. 8.1 The topology of the demonstrative example with three nodes, without self-feedback

Example 8.1 In our first example, the weight matrix W_1 is the following:

$$W_1 = \begin{bmatrix} 0 & 0.8 & -0.5 \\ 0.5 & 0 & 0.5 \\ 0.5 & -1 & 0 \end{bmatrix} \tag{8.17}$$

The iteration shows unstable behaviour, i.e. the activation values (coordinates of the activation vector) do not converge to a certain value, neither oscillate between several values (see Fig. 8.2). One may think that this strange behaviour is chaotic,

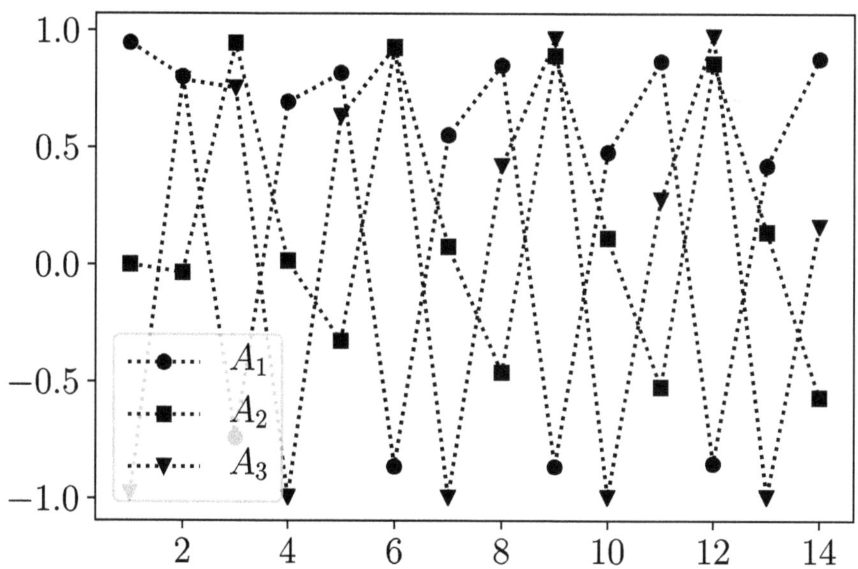

Fig. 8.2 FCM with hyperbolic tangent threshold function: unstable behaviour. If we look at the activation values coordinate-wise, we cannot see any trend to stabilization, so there is no straightforwrad pattern...

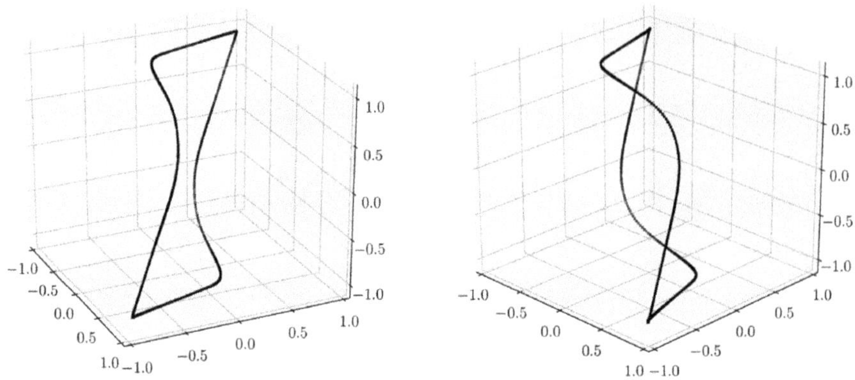

Fig. 8.3 ...But coupling coordinates together bring a new viewpoint: the points in the three dimensional space walk around a closed curve, this is the so-called quasiperiodic pattern. The left and right figures show the same curve in the three dimensional space from different viewpoints

but it is not the case. If we consider the three activation values as coordinates of point(s) in the three dimensional space, it turns out that activation vectors (i.e. the corresponding points) form a closed curve (see Fig. 8.3).

Example 8.2 In the second example, we introduce a small change in the weight matrix, namely we change the sign of w_{13}, everything else remains the same.

$$W_2 = \begin{bmatrix} 0 & 0.8 & 0.5 \\ 0.5 & 0 & 0.5 \\ 0.5 & -1 & 0 \end{bmatrix} \tag{8.18}$$

This relatively small change induces a qualitative change, now the iteration arrives to a limit cycle, it oscillates between two activation vectors (see Fig. 8.4). These vectors are symmetrical about the origin (the values are rounded to four decimal places):

$$A_{LC1} \begin{bmatrix} 0.5512 \\ 0 \\ -05512 \end{bmatrix} \quad \text{and} \quad A_{LC2} = \begin{bmatrix} -0.5512 \\ 0 \\ 05512 \end{bmatrix}. \tag{8.19}$$

Example 8.3 Finally, in the third example we apply only positive weights, the changes are $w_{13} = 0.5$ and $w_{32} = 1$.

$$W_3 = \begin{bmatrix} 0 & 0.8 & 0.5 \\ 0.5 & 0 & 0.5 \\ 0.5 & 1 & 0 \end{bmatrix} \tag{8.20}$$

The iteration leads to a fixed point (see Fig. 8.5).

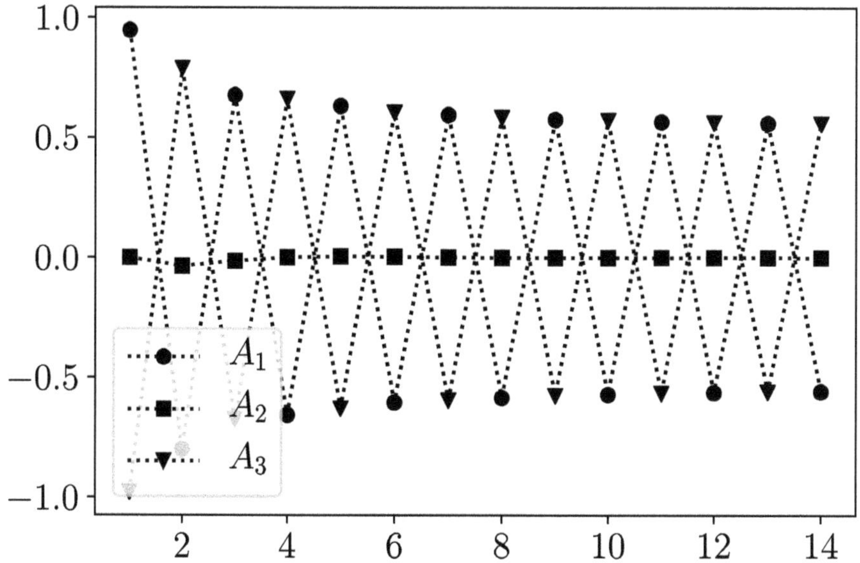

Fig. 8.4 FCM with hyperbolic tangent threshold function: limit cycle. Although the value of concept 2 is stabilized (converges to zero), the other two activation values keep oscillating

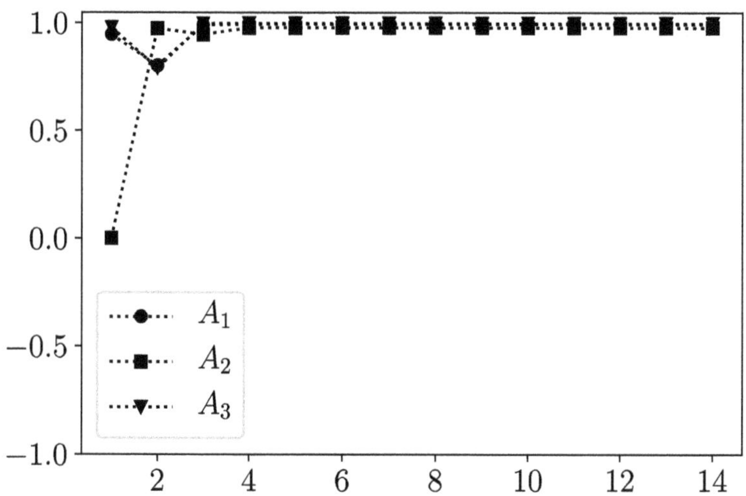

Fig. 8.5 FCM with hyperbolic tangent threshold function: fixed point. All the three activation values converge. The limit values are close to one, but slightly different from each other

The fixed point has the following coordinates (rounded to four decimals):

$$A_{final} = \begin{bmatrix} 0.9937 \\ 0.9776 \\ 0.9938 \end{bmatrix} \quad (8.21)$$

8.4 Summary and Mathematical Problems

As it can be seen from the previous examples and known from the literature, fuzzy cognitive maps with continuous threshold functions can show a large variety of behaviour. The behaviour (i.e. the limiting behaviour of the iteration) may depend on the following factors:

- Topology of the network;
- Weights assigned to the edges (weight matrix);
- Updating rule (existence, absence, strength of self-feedback);
- Parameter(s) of the threshold function(s);
- Initial concept vector.

From the application point of view, convergence to a fixed point is the desired behaviour of an FCM. Unfortunatelly, in many cases it is not the case, so the FCM based modelling usually follows the trial and error method. Consequently, there is an existing need for methods and theoretical findings, that are able to give support to the FCM based modelling.

In the following chapters, we introduce some result regarding the dynamics of fuzzy cognitive maps. Moreover, we discuss the existence and uniqueness of fixed points for extension of FCMs, like fuzzy set valued and fuzzy grey cognitive maps.

Acknowledgements This work was supported by the project 'Developing personalized insurance products using artificial intelligence', no. 2020-1.1.2-PIACI-KFI-2021-00267, National Research, Development and Innovation Office (NKFIH).

References

1. Apostolopoulos ID, Groumpos PP (2020) Non-invasive modelling methodology for the diagnosis of coronary artery disease using fuzzy cognitive maps. Comput Methods Biomech Biomed Eng 1–9
2. Groumpos PP (2010) Fuzzy cognitive maps: basic theories and their application to complex systems. In: Fuzzy cognitive maps. Springer, pp 1–22
3. Harmati IÁ, Hatwágner MF, Kóczy LT (2018) On the existence and uniqueness of fixed points of fuzzy cognitive maps. In: International conference on information processing and management of uncertainty in knowledge-based systems. Springer, pp 490–500
4. Harmati IÁ, Kóczy LT (2019) Notes on the dynamics of hyperbolic tangent fuzzy cognitive maps. In: 2019 IEEE international conference on fuzzy systems (FUZZ-IEEE). IEEE, pp 1–6

5. Kosko B (1993) Adaptive inference in fuzzy knowledge networks. In: Readings in fuzzy sets for intelligent systems. Elsevier, pp 888–891
6. Kosko B et al (1986) Fuzzy cognitive maps. Int J Man-Mach Stud 24(1):65–75
7. Kuznetsov YA (2013) Elements of applied bifurcation theory, vol 112. Springer Science & Business Media
8. Lazzerini B, Mkrtchyan L (2011) Analyzing risk impact factors using extended fuzzy cognitive maps. IEEE Syst J 5(2):288–297
9. Nápoles G, Papageorgiou E, Bello R, Vanhoof K (2016) On the convergence of sigmoid fuzzy cognitive maps. Inf Sci 349–350:154–171
10. Nápoles G, Papageorgiou E, Bello R, Vanhoof K (2017) Learning and convergence of fuzzy cognitive maps used in pattern recognition. Neural Process Lett 45(2):431–444
11. Papageorgiou EI, Salmeron JL (2012) A review of fuzzy cognitive maps research during the last decade. IEEE Trans Fuzzy Syst 21(1):66–79
12. Shojaei P, Haeri SAS (2019) Development of supply chain risk management approaches for construction projects: a grounded theory approach. Comput Ind Eng 128:837–850
13. Tsadiras AK (2008) Comparing the inference capabilities of binary, trivalent and sigmoid fuzzy cognitive maps. Inf Sci 178(20):3880–3894
14. Tsadiras AK (2007) Inference using binary, trivalent and sigmoid fuzzy cognitive maps. In: Proceedings of the 10th international conference on engineering applications of neural networks (EANN 2007), vol 284, pp 327–334
15. Wiggins S (2003) Introduction to applied nonlinear dynamical systems and chaos, vol 2. Springer Science & Business Media

Chapter 9
Dynamics of Classical Fuzzy Cognitive Maps

István Á. Harmati

9.1 Motivation and Preliminary Results

The existence and uniqueness of fixed point of sigmoid fuzzy cognitive maps was firstly discussed by Boutalis et al. in [1] for the case when the parameter of the log-sigmoid threshold function is $\lambda = 1$, (so the function was $f(x) = 1/(1 + e^{-x})$. Namely, they have proved that if the inequality $\left(\sum_{i=1}^{n} \sum_{j=1}^{n} w_{ij}^2\right)^{1/2} < 4$ holds, then the FCM with sigmoid threshold function of parameter value $\lambda = 1$ has one and only one fixed point. This result has been generalized for arbitrary $\lambda > 0$ steepness parameter in [3]. The behaviour of FCMs with linear and sigmoid transfer functions has been analyzed in [5]. It has been shown, that if the parameter of the sigmoid transfer function is small enough, then the FCM has exactly one fixed point. Nevertheless, this result is rather theoretical, since the weights are not taken into consideration. The stability of fuzzy cognitive maps was examined via Lyapunov stability analysis in [9] and was applied in clinical decision making [8]. The global stabilty of FCMs and the comparison of the existing results for unique fixed points is discussed in [4].

Example 9.1 The following simple toy example shows the qualitative change of the behaviour of a fuzzy cognitive map, when the parameter of the sigmoid threshold function changes. The topology of the map is shown in Fig. 9.1 and the long-time behaviour of the FCM is shown in Fig. 9.2.

As we have seen previously, after formalizing the updating rule of an FCM, it is obvious, that from the mathematical point of view, it is a discrete dynamical system. The dynamics of a discrete dynamical system may change dramatically, if its parameters are varied. A qualitative change of the dynamical behaviour, e.g. a transition from a unique stable fixed point to a completely different behaviour (multiple fixed points, limit cycle, quasi periodic motion), is called bifurcation and the corresponding critical value of the parameter is the so-called bifurcation point.

L. T. Kóczy, *Fuzzy Cognitive Maps*, Studies in Fuzziness and Soft Computing 427,
https://doi.org/10.1007/978-3-031-37959-8_9

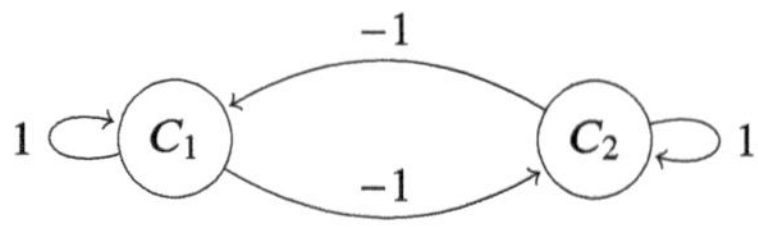

Fig. 9.1 The topology of the demonstrative example. The self-loops indicate the self-feedback

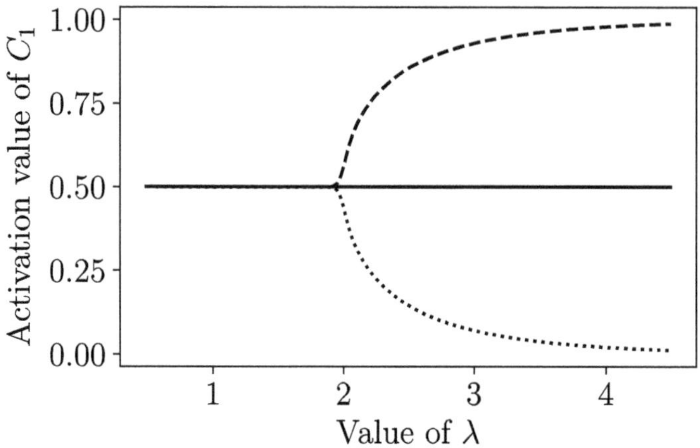

Fig. 9.2 The point [0.5, 0.5] is always a fixed point in the example FCM. Moreover, if $\lambda \le 2$, then it is the only fixed point and every initial stimulus leads to this point. But if the value of λ is greater than 2, then two other fixed points arises

To understand the dynamics of fuzzy cognitive maps, we need some elements of bifurcation theory. There are many comprehensive books with more detailed explanations, see for example [7] or [12].

Consider a discrete-time dynamical system depending on only one parameter λ ($G\colon \mathbb{R}^n \to \mathbb{R}^n$):

$$A(k+1) = G(A(k), \lambda), \qquad A(k) \in \mathbb{R}^n, \quad \lambda \in \mathbb{R}, \tag{9.1}$$

where function G is smooth with respect to A and λ. Let us assume that A^* is a fixed point of the mapping with parameter λ_0. The local stability (resistance against small perturbations) of the fixed point depends on the eigenvalues of the Jacobian evaluated at A^* ($J_G(A^*)$). If the Jacobian has no eigenvalues on the unit circle, A^* is said to be a *hyperbolic* fixed point (it has nothing to do with the *hyperbolic* threshold function). Hyperbolic fixed points can be categorized according to the eigenvalues of $J_G(A^*)$:

- If every eigenvalue falls within the unit circle (i.e., their absolute value is less than one), then it is an asymptotically stable fixed point. It means that this fixed point attracts the points of the space in every direction in its (sometimes very small) neighborhood.

- If there is at least one eigenvalue with absolute value greater than one, and there is at least one eigenvalue with absolute value less than one, then it is a saddle point. This fixed point may attract points of the space in some, but not in every direction in its neighborhood.
- If every eigenvalue has absolute value greater than one, then it is an unstable fixed point.

The following three cases are the simplest ways, when the hyperbolicity of the fixed point can be lost (α denotes an eigenvalue of the Jacobian evaluated at the fixed point):

- A simple positive eigenvalue crosses the unit circle at $\alpha = 1$. The bifurcation associated to this scenario is called saddle-node bifurcation. It is a birth of new fixed points.
- A simple negative eigenvalue crosses the unit circle at $\alpha = -1$. This is a period-doubling bifurcation, the birth of limit cycles.
- A pair of complex conjugate eigenvalues approaches the unit circle at $\alpha_{1,2} = e^{i\varphi}$, $\varphi \in (0, \pi)$. This is the so-called Neimark-Sacker bifurcation, causes the occurence of a closed invariant curve (quasiperiodic behaviour).

Example 9.2 In the previous chapter, we demonstrated the different behaviour of fuzzy cognitive maps. In those examples, the threshold function was the hyperbolic tangent function. The eigenvalues of the Jacobian evaluated at the trivial fixed point (zero) are just the eigenvalues of the weight matrix W multiplied by the parameter λ. In Table 9.1, we summarize the weight matrices, eigenvalues of the Jacobian at the zero fixed point, their absolute values and the type of bifurcations.

Table 9.1 Examples of weight matrices, eigenvalues and bifurcation

Weight matrix	Eigenvalues	Absolute values	Type of bifurcation
$W_1 = \begin{bmatrix} 0 & 0.8 & -0.5 \\ 0.5 & 0 & 0.5 \\ 0.5 & -1 & 0 \end{bmatrix}$	0.6144 $-0.3082 + 0.7968\,i$ $-0.3082 - 0.7968\,i$	0.6144 0.8544 0.8544	Quasiperiodic
$W_2 = \begin{bmatrix} 0 & 0.8 & 0.5 \\ 0.5 & 0 & 0.5 \\ 0.5 & -1 & 0 \end{bmatrix}$	-0.5 $0.25 + 0.1936\,i$ $0.25 - 0.1936\,i$	0.5 0.3162 0.3162	Limit cycle
$W_3 = \begin{bmatrix} 0 & 0.8 & -0.5 \\ 0.5 & 0 & 0.5 \\ 0.5 & 1 & 0 \end{bmatrix}$	1.2311 -0.5 -0.7311	1.2311 0.5 0.7311	New fixed points

9.2 Mathematical Background

Before starting the mathematical discussion, we give a much more compact and general form of the updating rule. Kosko's original updating rule was the following [6]:

$$A_i(k+1) = f\left(\sum_{j=1, j\neq i}^{n} A_j(k) \cdot w_{ij}\right) \tag{9.2}$$

Later, another updating rule has been introduced, taking into account the current state of the concept itself (self-feedback) [2]:

$$A_i(k+1) = f\left(A_i(k) + \sum_{j=1, j\neq i}^{n} A_j(k) \cdot w_{ij}\right) \tag{9.3}$$

A more general form, when the effect of self-feedback is taken into consideration with a factor, expressing the strength of memory ($0 \leq d_i \leq 1$):

$$A_i(k+1) = f\left(d_i A_i(k) + \sum_{j=1, j\neq i}^{n} A_j(k) \cdot w_{ij}\right) \tag{9.4}$$

Now let us introduce the following notation:

$$w_{ij}^* = \begin{cases} w_{ij} & \text{if } i \neq j \\ d_i & \text{if } i = j \end{cases} \tag{9.5}$$

Using this notation we have

$$\begin{aligned} A_i(k+1) &= f\left(d_i A_i(k) + \sum_{j=1, j\neq i}^{n} A_j(k) \cdot w_{ij}\right) = f\left(A_i(k) \cdot w_{ii}^* + \sum_{j=1, j\neq i}^{n} A_j(k) \cdot w_{ij}^*\right) \\ &= f\left(\sum_{j=1}^{n} A_j(k) \cdot W_{ij}^*\right) = f\left(w_i^* A(k)\right), \end{aligned} \tag{9.6}$$

where w_i^* is the ith row of W^* and $A(k)$ is the concept vector at time instant k. In vectorial form, the equation becomes the following:

$$A(k+1) = f\left(W^* A(k)\right), \tag{9.7}$$

where function f is applied coordinate-wise. Moreover,

$$W^* = W + \text{diag}[d_i], \tag{9.8}$$

where diag$[d_i]$ is a diagonal matrix with d_is in its diagonal and zeros elsewhere, W is the original weight matrix, which contains zeros in the main diagonal. Note that if $d_i = 0$ for every $i = 1 \ldots n$, then $W^* = W$, which yields Kosko's model. If $d_i = 1$ for every $i = 1 \ldots n$, then we have $W^* = W + I$, which is the model described by Eq. 9.3.

The *existence* of a fixed point is clear from the following well-known theorems of topology:

Theorem 9.1 (Brouwer's fixed point theorem) *Every continuous function from a convex compact subset K of a Euclidean space $\mathbb{R}^n$ to K itself has at least one fixed point.*

Or a more general statement:

Theorem 9.2 (Schauder's fixed point theorem) *Every continuous function from a convex compact subset K of a Banach space to K itself has at least one fixed point.*

Nevertheless, these theorems prove only the *existence* of at least one fixed point for any fuzzy cognitive map with continuous threshold function, and do not provide any information how we may find fixed points. Since the FCM reasoning is based on the limit of the iteration, we may wonder under what conditions this limit does exist. Intuitively, if the distance between two consecutive concept vectors in the iteration becomes smaller and smaller, the iteration converges. From the mathematical point of view, this situation can be described by the so-called *contraction*.

Definition 9.1 ([11]) Let (X, d) be a metric space. A mapping $f : X \to X$ is a *contraction mapping* or *contraction* if there exists a constant c (independent from x and y), with $0 \leq c < 1$, such that

$$d\left(f(x), f(y)\right) \leq cd(x, y) \tag{9.9}$$

The iteration process of an FCM stops when the concept vector does not change anymore (technically it means that the difference between two consecutive iterations is less than a predefined small value), so the concept vector is a fixed point in the n dimensional space. More precisely, a fixed point of mapping $f : X \to X$ is a point $x^* \in X$ such that $f(x^*) = x^*$.

The following theorem ensures the existence and uniqueness of a fixed point of a contraction mapping. Moreover, it provides an iterative algorithm to determine the fixed point.

Theorem 9.3 (Banach's fixed point theorem) *If $f : X \to X$ is a contraction mapping on a nonempty complete metric space (X, d), then f has only one fixed point x^*. Moreover, x^* can be found as follows: start with an arbitrary $x_0 \in X$ and define the sequence $x_n = f(x_{n-1})$, then $\lim\limits_{n\to\infty} x_n = x^*$.*

9.3 The Sigmoid Case

The most widely used threshold function in applications of FCMs is the sigmoid function. In the proof of the theorem regarding the existence and uniqueness of fixed points of FCMs with sigmoid threshold function, the following (maybe well-known) fact plays an important role.

Lemma 9.1 *The derivative of the log-sigmoid function $f\colon \mathbb{R} \to \mathbb{R}$, $f(x) = 1/(1 + e^{-\lambda x})$ ($\lambda > 0$) is bounded by $\lambda/4$.*

Proof The derivative of the sigmoid function is:

$$f'(x) = \left(\frac{1}{1 + e^{-\lambda x}}\right)' = \frac{\lambda e^{-\lambda x}}{(1 + e^{-\lambda x})^2} = \lambda \cdot f(x) \cdot (1 - f(x)) \tag{9.10}$$

Since $0 < f(x) < 1$ for every $x \in \mathbb{R}$, the inequality $0 < f(x)(1 - f(x)) \le 1/4$ holds for every $x \in \mathbb{R}$, which implies that $0 < f'(x) \le \lambda/4$.

The following theorem is a special case of the results presented in [4] but, since sigmoid functions are the most widely used threshold functions for FCMs, it is useful to discuss this case separately.

Theorem 9.4 *Consider a fuzzy cognitive map (FCM) with weight matrix W. Moreover, let λ_i be the parameter of the sigmoid threshold function f_i applied at the ith concept. If the inequality*

$$\|\mathrm{diag}(\lambda_i/4) \cdot W\|_* < 1 \tag{9.11}$$

holds with an induced matrix norm $\|\cdot\|_*$, then the fuzzy cognitive map has the same fixed point for every initial activation vector.

Proof According to our previous notation, we have

$$G(A) = [f_1(w_1 A), f_2(w_2 A), \ldots, f_n(w_n A)]^T \tag{9.12}$$

where $w_i = (w_{i1}, \ldots, w_{in})$. Moreover, let $\|\cdot\|_v$ be the vector norm generating matrix norm $\|\cdot\|_*$ (i.e. for matrix M, $\|M\|_* = \sup\left\{\frac{\|Mx\|_v}{\|x\|_v},\ x \neq 0\right\}$. In the following, we give an upper estimation of the value of $\left\|G(A) - G(A')\right\|_v$:

$$\left\|G(A) - G(A')\right\|_v = \left\|\left[f_1(w_1 A) - f_1(w_1 A'), \ldots, f_n(w_n A) - f_n(w_n A')\right]\right\|_v \tag{9.13}$$

From the Mean Value Theorem, we get the following upper bound for the coordinate-wise difference:

$$\left|f_i(w_i A) - f_i(w_i A')\right| \le \frac{\lambda_i}{4}\left|w_i A - w_i A'\right|. \tag{9.14}$$

So we have

$$\left\|G(A) - G(A')\right\|_v \leq \left\| \left[\frac{\lambda_1}{4}(w_1 A - w_1 A'), \ldots, \frac{\lambda_n}{4}(w_n A - w_n A') \right]^T \right\|_v \tag{9.15}$$

$$= \left\| \operatorname{diag}(\frac{\lambda_i}{4} K_i) W (A - A') \right\|_v \tag{9.16}$$

$$= \frac{\left\| \operatorname{diag}(\frac{\lambda_i}{4}) W (A - A') \right\|_v}{\|A - A'\|_v} \cdot \left\|A - A'\right\|_v \tag{9.17}$$

$$\leq \left\| \operatorname{diag}(\frac{\lambda_i}{4}) W \right\|_* \cdot \left\|A - A'\right\|_v, \tag{9.18}$$

where the last row is the consequence of the definition of matrix norm $\|\cdot\|_*$. If $\left\|\operatorname{diag}(\frac{\lambda_i}{4})W\right\|_* < 1$, then G is a contraction mapping. It means that starting from an arbitrary initial concept vector, the iteration leads to the same equilibrium point. ■

We should note that this is a sufficient, but not necessary condition. The fact that the inequality $\left\|\operatorname{diag}(\frac{\lambda_i}{4})W\right\|_* < 1$ holds implies that there is one and only one fixed point. Nevertheless, if $\left\|\operatorname{diag}(\frac{\lambda_i}{4})W\right\|_* > 1$, we cannot ensure that there are more than one fixed points or other complex patterns occur.

9.4 Hyperbolic Tangent Case

In some cases, the hyperbolic tangent function is applied as a threshold function, which squashes the concept values into the interval $[-1, 1]$. Before stating the theorem regarding the dynamics of hyperbolic FCMs, take a look at the hyperbolic tangent function in one dimension (see Fig. 9.3). It will bring some useful analogies and intuition about the dynamics in higher dimensional spaces. The following facts can be easily checked using basic calculus:

- Zero is always a fixed point (trivial fixed point).
- If the slope parameter λ is less than one, then zero is the unique fixed point. In this case, every iteration leads to this point, regardless of the initial value.
- If the slope parameter is greater than one, then there are two other fixed points, symmetrical about zero. In this case, if the starting value $x_0 < 0$, then the iteration leads to the left fixed point; while if $x_0 > 0$, then the iteration leads to the right fixed point. Only $x_0 = 0$ yields the trivial fixed point.

As we will see, a very similar pattern can be observed in the higher dimensional case.

It is easy to see, that a similar statement can be formalized as in Theorem 9.4. Nevertheless, there is a great difference between the sigmoid and the hyperbolic tangent FCMs: in the latter case, there is a *known* fixed point. This fact allows us to provide a stronger statement. First, we state the bound of the derivative of the threshold function.

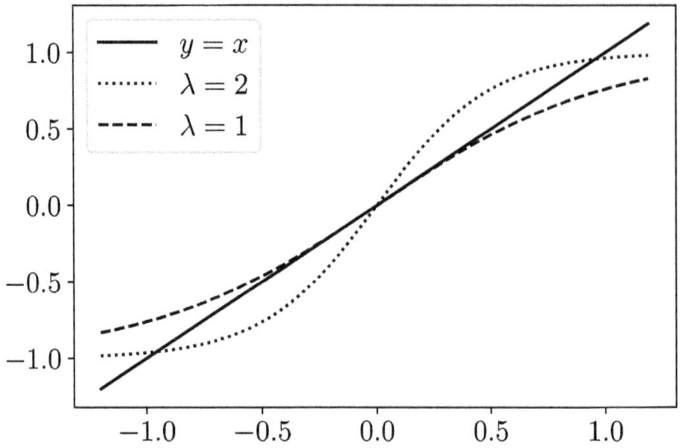

Fig. 9.3 Hyperbolic tangent functions with different steepness parameter λ. We can observe that zero is always a fixed point and for larger value of λ two other fixed points appear, which are symmetrical about the trivial fixed point 0

Lemma 9.2 *The derivative of the hyperbolic tangent function* $f : \mathbb{R} \to \mathbb{R}$, $f(x) = \dfrac{e^{2\lambda x} - 1}{e^{2\lambda x} + 1}$ *(*$\lambda > 0$*) is bounded by* λ*.*

Proof The derivative of $f(x)$ is

$$f'(x) = \frac{4\lambda e^{2\lambda x}}{\left(e^{2\lambda x} + 1\right)^2} = \lambda(1 - f^2(x))$$

Since $-1 < f(x) < 1$, it follows that $0 < f'(x) < \lambda$.

We know that if the spectral radius of the Jacobian at a fixed point is less than one, then this point is a fixed point attractor; if the spectral radius at this point is more than one, then this point is a repellor state (at least in some directions). For further analysis, we compute the Jacobian of mapping G.

An arbitrary entry of the Jacobian of mapping G:

$$\frac{\partial G_i}{\partial A_j} = \lambda w_{ij}(1 - f^2(w_i A)) \tag{9.19}$$

Since

$$f(w_i A)\big|_{A=[0,\ldots,0]^T} = 0 \tag{9.20}$$

the value of this expression at $A^* = [0, \ldots, 0]^T$ is $\lambda \cdot w_{ij}$. Consequently, the Jacobian of mapping G at the known fixed point $A^* = [0, \ldots, 0]^T$ is

$$J_G = \lambda \begin{bmatrix} w_{11} & w_{12} & \dots & w_{1n} \\ w_{21} & w_{22} & \dots & w_{2n} \\ \vdots & \vdots & \ddots & \vdots \\ w_{n1} & w_{n2} & \dots & w_{nn} \end{bmatrix} = \lambda W \tag{9.21}$$

If the spectral radius of the Jacobian at a fixed point is less than one, then there exists a matrix norm, such that the mapping is a contraction w.r.t. this norm in the neighbourhood of this fixed point, so this fixed point is asymptotically stable. Moreover, since $0 < 1 - f^2(w_i A) \leq 1$, the norm of the Jacobian at any other point is less than or equals to the norm at $A = [0, \dots, 0]^T$. This implies that the norm of the Jacobian at the zero fixed point is an upper bound for the norm of Jacobians for the entire space. If $\|\lambda W\| < 1$ w.r.t. any matrix norm, then mapping G is a contraction on the entire space, so it has one and only one fixed point.

Summarizing this reasoning, we can conclude the following theorem:

Theorem 9.5 *Let W be the weight matrix (including possible feedback) of an FCM and let $\lambda > 0$ be the parameter of the hyperbolic tangent threshold function. Let us denote by $\rho(W)$ the spectral radius of W. Then the following statements hold:*

- *if $\rho(W) < 1/\lambda$ (i.e. $\lambda < 1/\rho(W)$), then $A^* = [0, \dots, 0]^T$ is a fixed point attractor and its basin of attraction is the whole space $\mathbb{R}^n$. The iteration will end up at this point, regardless of the initial value.*
- *if $\rho(W) > 1/\lambda$ (i.e. $\lambda > 1/\rho(W)$), then $A^* = [0, \dots, 0]^T$ is a fixed point, but not a global attractor. There exist initial concept vectors such that the iteration does not lead to the trivial $[0, \dots, 0]^T$ fixed point.*

We observed in the one dimensional case that if x^* is a fixed point of $f(x) = \tanh(\lambda x)$, then $-x^*$ is also a fixed point. Similar statement holds for the fixed points of hyperbolic tangent FCMs. The following theorem shows that the fixed points of a hyperbolic tangent FCM are given in pairs:

Theorem 9.6 *If A^* is a fixed point of a hyperbolic tangent fuzzy cognitive map, then $-A^*$ is also a fixed point of this FCM.*

Proof Let us assume that $A^* = [A_1^*, \dots, A_n^*]^T$ is a fixed point of the hyperbolic tangent FCM, so $G(A^*) = A^*$. Then

$$G(-A^*) = [f(w_1(-A^*)), \dots, f(w_n(-A^*))]^T \tag{9.22}$$

Applying the fact that f is an odd function we get that

$$G(-A^*) = [-f(w_1 A^*), \dots, -f(w_n A^*)]^T = -[f(w_1 A^*), \dots, f(w_n A^*)]^T \tag{9.23}$$

$$= -[A_1^*, \dots, A_n^*]^T = -A^* \tag{9.24}$$

which means that $-A^*$ is a fixed point. □

We have seen that the fixed points of a hyperbolic tangent FCM have a symmetric structure. The similar statement is true for their basins of attraction (basin of attarction: the set of initial concept vectors, from which the iteration tends to a certain fixed point).

Let A^* be a fixed point of a hyperbolic FCM. Let us denote its basin of attraction by H:

$$H = \left\{ A \in \mathbb{R}^n : \lim_{n\to\infty} A(k)\Big|_{A(0)=A} = A^* \right\} \tag{9.25}$$

Theorem 9.7 *If A^* is a fixed point of a hyperbolic fuzzy cognitive map, and its basin of attraction is $H \subset \mathbb{R}^n$, then the basin of attraction of the fixed point $-A^*$ is the set*

$$H' = \left\{ -A \in \mathbb{R}^n : A \in H \right\} \tag{9.26}$$

Proof Let us assume that $A \in \mathbb{R}^n$ is an element of H, so if the iteration starts from this point then by iteration we get

$$A(0) = A \tag{9.27}$$
$$A(1) = G(A(0)) \tag{9.28}$$
$$A(2) = G(A(1)) \tag{9.29}$$
$$\vdots \tag{9.30}$$

Since $A \in H$, then

$$\lim_{k\to\infty} A(k) = A^* \tag{9.31}$$

We are going to show that if the iteration starts from $A'(0) = -A$, then the limit is the fixed point $-A^*$. The iteration is generated by the rule $A(k+1) = G(A(k))$, so if we start with $-A$, then

$$A'(1) = G(A'(0)) = G(-A) = [f(w_1(-A)), \ldots, f(w_n(-A))]^T \tag{9.32}$$
$$= [-f(w_1 A), \ldots, -f(w_n A)]^T = -[f(w_1 A), \ldots, f(w_n A)]^T \tag{9.33}$$
$$= -[A_1, \ldots, A_n]^T = -A(1) \tag{9.34}$$

Repeating the process above, we get that the consecutive elements of the iteration are

$$A'(1) = -A(1) \tag{9.35}$$
$$A'(2) = -A(2) \tag{9.36}$$
$$A'(3) = -A(3) \tag{9.37}$$
$$\vdots \tag{9.38}$$

It is known that $\lim_{k\to\infty} A(k) = A^*$, so we have

$$\lim_{k\to\infty} A'(k) = \lim_{k\to\infty} -A(k) = -A^* \tag{9.39}$$

which means that $-A$ belongs to the basin of attraction of fixed point $-A^*$. Since the same proof works for every $A \in H$, we can conlude that $H' = -H$ (i.e. if $A \in H$, then $-A \in H'$).

Consider the previous theorem for the case of limit cycle. Assume that the iteration started from the point $A(0) = A$ results in a limit cycle whose elements are $A^{1*}, \dots, A^{d*}$. Following a similar mathematical argument that was applied in the proof of Theorem 9.6 and Theorem 9.7 , it follows that if the iteration starts with $A(0) = -A$, then it leads to a limit cycle with elements $-A^{1*}, \dots, -A^{d*}$. Note that the sets $\{A^{1*}, \dots, A^{d*}\}$ and $\{-A^{1*}, \dots, -A^{d*}\}$ are not necesseraly different.

9.5 The Rescaled Sigmoid Case

The FCM based decision-making usually assigns zero activation value to the nodes whose values are unknown. A new, modified updating rule was proposed in [10] for the cases where we have no sufficient information about the value of the concept or the human expert can not provide efficient estimation of this.

The updating rule is the following:

$$A_i(k+1) = f\left(\sum_{j=1, j\neq i}^{n} w_{ij}(2A_j(k) - 1) + 2A_i(k) - 1\right) \tag{9.40}$$

Similarly to the updating rule of the original FCM reasoning, this one also has a more compact form. If we include ones into the diagonal of the weight matrix W, then the updating rule above can be rewritten in the following form:

$$A(k+1) = f(W(2A(k) - 1) \tag{9.41}$$

where 1 means a vector with ones and the sigmoid function f understood coordinate-wise. The updating process is again a mapping $G\colon \mathbb{R}^n \to \mathbb{R}^n$, where $G(A) = [f(w_1(2A(k) - 1), \dots, f(w_n(2A(k) - 1)]^T$ and w_i is the ith row of W (including ones in its main diagonal).

It is obvious that $A = [0.5, \dots, 0.5]^T$ is a fixed point of mapping G, since if $A = [0.5, \dots, 0.5]^T$, then $2A - 1 = [0, \dots, 0]^T$, so $w_i(2A(k) - 1) = 0$ and

$$G(A)_i = f(w_i(2A - 1)) = \frac{1}{1 + e^{-\lambda\cdot 0}} = 0.5. \tag{9.42}$$

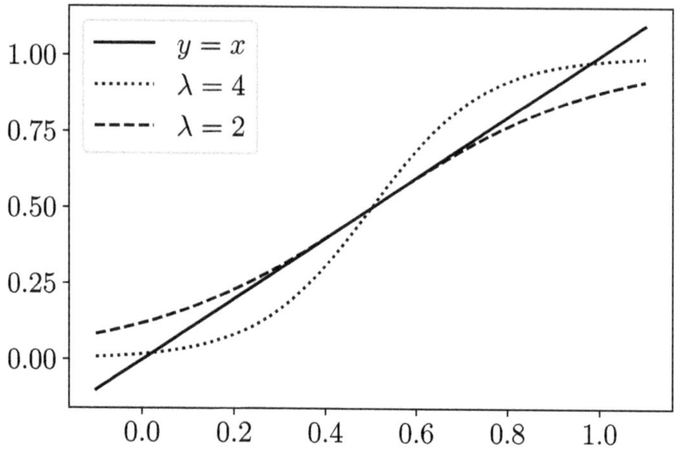

Fig. 9.4 Rescaled sigmoid functions for different values of parameter λ. We can observe that $1/2$ is always a fixed point and for larger value of λ two other fixed points appear, which are symmetrical about the trivial fixed point $1/2$.The similarity to the hyperbolic tangent case is obvious

It implies that $G([0.5, \ldots, 0.5]^T) = [0.5, \ldots, 0.5]^T$ regardless of the value of parameter λ and regardless of the weight matrix W.

There is an obvious similarity between hyperbolic tangent and rescaled sigmoid FCMs (see Figs. 9.3 and 9.4). Both of them have trivial fixed points and their threshold functions are symmetrical about a point, which the trivial fixed point in one dimension.

Similarly to the hyperbolic tangent case, we determine the Jacobian of the updating mapping at the trivial fixed point. A general element of the Jacobian matrix is

$$\frac{\partial f_i}{\partial A_j} = 2\lambda w_{ij} f(w_i(2A-1)) \cdot (1 - f(w_i(2A-1))) \tag{9.43}$$

Since $f(w_i(2A-1))\big|_{A=[0.5,\ldots,0.5]^T} = 0.5$, the value of this expression at $A^* = [0.5, \ldots, 0.5]^T$ is $\lambda/2 \cdot w_{ij}$. So the Jacobian of mapping G at the known fixed point A^* is

$$J_G = \frac{\lambda}{2}\begin{bmatrix} w_{11} & w_{12} & \ldots & w_{1n} \\ w_{21} & w_{22} & \ldots & w_{2n} \\ \vdots & \vdots & \ddots & \vdots \\ w_{n1} & w_{n2} & \ldots & w_{nn} \end{bmatrix} = \frac{\lambda}{2} W \tag{9.44}$$

where the elements of the main diagonal are $w_{11} = \ldots = w_{nn} = 1$. We know that if the norm of the Jacobian matrix at the fixed point is less than one, then the mapping is a contraction at this point, so this fixed point is asymptotically stable. Since $f(w_i(2A - 1)) \cdot (1 - f(w_i(2A-1))) \leq 1/4$ we got that the norm of the Jacobian at any other point is at most the norm at $A = [0.5, \ldots, 0.5]^T$, so it is an upper bound for the entire space. If $\|\frac{\lambda}{2}W\| < 1$ w.r.t. any natural matrix norm, then mapping G is a contraction on the entire space, so it has one and only one fixed point.

Summarizing it, we get the following statement:

Theorem 9.8 *Let W be the extended weight matrix (including ones in the diagonal) of an FCM and let $\lambda > 0$ be the parameter of the sigmoid threshold function. The updating rule is implemented by the rescaled mapping: $A^{(k+1)} = f(W(2A^{(k)} - 1)$. Let us denote by $\rho(W)$ the spectral radius of W. Then the following statements hold:*

- *if $\rho(W) < 2/\lambda$ (i.e. $\lambda < 2/\rho(W)$), then $A^* = [0.5, \ldots, 0.5]^T$ is a global fixed point attractor. The iteration leads to this point, regardless of the initial concept vector.*
- *if $\rho(W) > 2/\lambda$ (i.e. $\lambda > 2/\rho(W)$), then $A^* = [0.5, \ldots, 0.5]^T$ is a fixed point, but not global attractor. There exist initial concept vectors, such that the iteration does not lead to this fixed point.*

Similarly to the hyperbolic tangent case, the fixed points of rescaled sigmoid FCMs are symmetrical.

Theorem 9.9 *If A^* is a fixed point of a rescaled sigmoid fuzzy cognitive map, then $1 - A^*$ is also a fixed point of it.*

Proof In the proof, we apply the compact formalism. Since A^* is a fixed point, the equality $f(W(2A^* - 1)) = A^*$ holds. Then using the fact that for any sigmoid function $f(-x) = 1 - f(x)$, we get that

$$\begin{aligned} G(1 - A^*) &= f(W(2(1 - A^*) - 1)) = f(W(1 - 2A^*) = 1 - f(W(2A^* - 1)) \\ &= 1 - A^* \end{aligned} \tag{9.45}$$

I.e. $G(1 - A^*) = 1 - A^*$, which means that $1 - A^*$ is also a fixed point.

Similar statement holds for the basins of attractions of fixed points.

Theorem 9.10 *If A^* is a fixed point of a rescaled sigmoid fuzzy cognitive map, and its basin of attraction is $H \subset \mathbb{R}^n$, then the basin of attraction of the fixed point $1 - A^*$ is the set*

$$H' = \left\{1 - A \in \mathbb{R}^n : A \in H\right\} \tag{9.46}$$

Proof Let us assume that A belongs to the basin of attraction of A^*, so the series $A = A(0), A(1), A(2), \ldots$ converges to A^*. Then define $A' = 1 - A$.

$$A' = 1 - A \tag{9.47}$$

$$A'(1) = G(A') = G(1 - A) = f(2W(1 - A) - 1) = f(W(1 - 2A) \tag{9.48}$$

$$= 1 - f(W(2A - 1)) = 1 - A(1) \tag{9.49}$$

$$A'(2) = 1 - A(2) \tag{9.50}$$

$$\vdots \tag{9.51}$$

Since $\lim_{k\to\infty} A(k) = A^*$, then $\lim_{k\to\infty} A'(k) = 1 - A^*$, which means that $1 - A$ is in the basin of attraction of fixed point $1 - A^*$.

9.6 Summary

In the FCM based modelling methodology, the conclusion about the model is based on the iterative application of transformation (threshold) functions. In most of the cases, these function are sigmoid or hyperbolic tangent functions. The iteration may result in a fixed point, limit cycle or may produce chaotic behaviour.

In this chapter, theorems regarding of the existence and uniqueness of fixed points have been introduced for various fuzzy cognitive maps, using the weights of the interconnections in the FCM and the parameter of the threshold function. These results may be useful in the proper choice of the parameter of the threshold function applied, since in some cases the unique fixed point is the desired feature, but in most of the cases we prefer FCMs with not only one fixed points. The theoretical findings presented in this chapter may give a hint in both of these cases, i.e. what parameter values should be avoided or preferred.

Additionally, interesting symmetrical properties about the trivial fixed point of hyperbolic tangent and rescaled sigmoid fuzzy cognitive maps were also analyzed.

Acknowledgements This work was supported by the project 'Developing personalized insurance products using artificial intelligence', no. 2020-1.1.2-PIACI-KFI-2021-00267, National Research, Development and Innovation Office (NKFIH).

References

1. Boutalis Y, Kottas TL, Christodoulou M (2009) Adaptive estimation of fuzzy cognitive maps with proven stability and parameter convergence. IEEE Trans Fuzzy Syst 17(4):874–889
2. Groumpos PP (2010) Fuzzy cognitive maps: basic theories and their application to complex systems. In: Fuzzy cognitive maps. Springer, pp 1–22
3. Harmati IÁ, Hatwágner MF, Kóczy LT (2018) On the existence and uniqueness of fixed points of fuzzy cognitive maps. In: International conference on information processing and management of uncertainty in knowledge-based systems. Springer, pp 490–500
4. Harmati IÁ, Hatwágner MF, Kóczy LT (2021) Global stability of fuzzy cognitive maps. Neural Comput Appl 1–13
5. Knight CJK, Lloyd DJB, Penn AS (2014) An analysis of fixed points. Linear and sigmoidal fuzzy cognitive maps. Appl Soft Comput 15:193–202
6. Kosko B (1993) Adaptive inference in fuzzy knowledge networks. In: Readings in fuzzy sets for intelligent systems. Elsevier, pp 888–891
7. Kuznetsov YA (2013) Elements of applied bifurcation theory, vol 112. Springer Science & Business Media
8. Lee IK, Kim HS, Cho H (2012) Design of activation functions for inference of fuzzy cognitive maps: application to clinical decision making in diagnosis of pulmonary infection. Healthc. Inf. Res **18**(2):105–114

9. Lee I, Kwon SH (2010) Design of sigmoid activation functions for fuzzy cognitive maps via Lyapunov stability analysis. IEICE Trans Inf Syst 93(10):2883–2886
10. Papageorgiou EI (2011) A new methodology for decisions in medical informatics using fuzzy cognitive maps based on fuzzy rule-extraction techniques. Appl Soft Comput 11(1):500–513
11. Rudin W et al (1964) Principles of mathematical analysis, vol 3. McGraw-Hill, New York
12. Wiggins S (2003) Introduction to applied nonlinear dynamical systems and chaos, vol 2. Springer Science & Business Media

Chapter 10
Dynamics of Fuzzy Cognitive Maps with Uncertain Weights

István Á. Harmati

10.1 Introduction

The experts' knowledge or opinion in real-life problems are rarely exact, the rather provide approximations or estimates with or without margin of error. The weights of the connections in a fuzzy cognitive map model are usually based on experts' knowledge. Consequently, giving exact value to the weights hide the inherent uncertainty of human experts. To overcome this situation, an extension of FCMs has been introduced under the name of fuzzy grey cognitive maps (FGCMs) [9]. In this case, the concept vector is a vector of grey numbers, i.e. technically intervals.

Other approach assigns fuzzy sets to the weights, representing the most likely (according to the expert's opinion) value and the unknown range. In this case, the concept vector is a vector of fuzzy sets, usually fuzzy numbers.

10.2 Fuzzy Grey Cognitive Maps

In this section, we introduce sufficient conditions for the existence and uniqueness of fixed points of FGCMs. First, we shortly review the basic mathematical tools related to grey system theory.

10.2.1 Basic Operations on Grey Numbers

Fuzzy grey cognitive maps require numerical operations on the so-called grey numbers. A grey number (denoted by $\otimes g$) is a number whose accurate value is unknown, but the range of its possible values is known. A grey number with both a lower limit ($\underline{g}$) and an upper limit ($\overline{g}$) is called an interval grey number [5, 6], so $\otimes g \in [\underline{g}, \overline{g}]$.

L. T. Kóczy, *Fuzzy Cognitive Maps*, Studies in Fuzziness and Soft Computing 427,
https://doi.org/10.1007/978-3-031-37959-8_10

In applications, a grey number is usually an interval. The basic arithmetic operations on grey numbers are the following [6]:

1. $\otimes g_1 + \otimes g_2 \in [\underline{g_1} + \underline{g_2}, \overline{g_1} + \overline{g_2}]$
2. $- \otimes g \in [-\overline{g}, -\underline{g}]$
3. $\otimes g_1 - \otimes g_2 \in [\underline{g_1} - \overline{g_2}, \overline{g_1} - \underline{g_2}]$
4. $\otimes g_1 \times \otimes g_2 \in [\min(S), \max(S)]$,
   where $S = \left\{\underline{g_1} \cdot \underline{g_2}, \underline{g_1} \cdot \overline{g_2}, \overline{g_1} \cdot \underline{g_2}, \overline{g_1} \cdot \overline{g_2}\right\}$
5. If $\lambda > 0$, $\lambda \in \mathbb{R}$, then $\lambda \cdot \otimes g \in [\lambda \underline{g}, \lambda \overline{g}]$

Beside the above defined operations, we have to provide a consistent definition for the genaralization of any $f \colon \mathbb{R} \to \mathbb{R}$ function to grey numbers. The function of a grey number $\otimes g \in [\underline{g}, \overline{g}]$ is the grey number $f(\otimes g) \in [\underline{f(\otimes g)}, \overline{f(\otimes g)}]$, where

$$\underline{f(\otimes g)} = \inf\{f(\gamma) \colon \gamma \in [\underline{g}, \overline{g}]\} \tag{10.1}$$

$$\overline{f(\otimes g)} = \sup\{f(\gamma) \colon \gamma \in [\underline{g}, \overline{g}]\} \tag{10.2}$$

For a continuous and monotone increasing function f we have

$$\inf\{f(\gamma) \colon \gamma \in [\underline{g}, \overline{g}]\} = f(\underline{g}) \tag{10.3}$$

$$\sup\{f(\gamma) \colon \gamma \in [\underline{g}, \overline{g}]\} = f(\overline{g}) \tag{10.4}$$

Consequently, $\underline{f(\otimes g)} = f(\underline{g})$ and $\overline{f(\otimes g)} = f(\overline{g})$ and $f(\otimes g) \in [f(\underline{g}), f(\overline{g})]$.

Modelling via FGCM begins with the initial grey vector state $A(0)$, which represents initial uncertainty. The elements of this vector are grey numbers, i.e. $A_i(0) \in [\underline{A_i(0)}, \overline{A_i(0)}]$ for every i. The updated nodes' states are computed by an iterative process with an activation function, resulting grey numbers as concept values:

$$A_i(k) \in \left[f(\underline{w_i A(k-1)}), f(\overline{w_i A(k-1)})\right] \tag{10.5}$$

After a certain number of iterations, an FGCM with continuous threshold function arrives at one of the following cases:

1. It converges to a so-called grey fixed-point attractor. This fixed point is a vector, whose coordinates are intervals. The convergence (stabilization) means that the endpoints of these intervals are stabilized after a certain number of iterations.
2. The state keep cycling between several states, known as a limit grey cycle. These states (elements of the limit cycle) are grey concept vectors with interval coordinates. After a certain number of iteration steps, the elements of the limit cycle become stabilized, i.e. the system keeps cycling periodically over these finite number of state vectors.
3. The FGCM continues to produce different grey vector states for each iteration, this is the grey chaotic attractor.

10.2.2 Convergence of Fuzzy Grey Cognitive Maps

A fuzzy cognitive map with grey weights (FGCM) produces grey concept values, regardless of the nature of the initial concept vector. In the following, we assume that the human expert or the training process assigns the proper signs to the weights, so a weight is either positive or negative (more exactly nonnegative or nonpositive). This means that the type of relationship (direct or inverse) between the concepts is properly described by the fuzzy cognitive map, so in most of the cases, it seems to be a right assumption.

Let $\otimes w_{ij}$ be a weight describing the connection between concepts C_j and C_i. Due to our assumptions, $\otimes w_{ij}$ is a grey number and it is element of a subset of the interval $[-1, 0]$ or the interval $[0, 1]$, so

$$\otimes w_{ij} \in [\underline{w_{ij}}, \overline{w_{ij}}] \subseteq [-1, 0] \quad \text{or} \quad \otimes w_{ij} \in [\underline{w_{ij}}, \overline{w_{ij}}] \subseteq [0, 1] \tag{10.6}$$

Let us introduce the following notation:

$$w_{ij}^* = \begin{cases} |\underline{w_{ij}}| & \text{if } \otimes w_{ij} \leq 0 \\ \overline{w_{ij}} & \text{if } \otimes w_{ij} \geq 0 \end{cases} \tag{10.7}$$

In other words, w_{ij}^* is the absolute value of the most extreme value of $\otimes w_{ij}$ possible (it equals the absolute value of the interval $[\underline{w_{ij}}, \overline{w_{ij}}]$).

In the following, we provide sufficient conditions for the existence and uniqueness of fixed point of FGCMs [3, 4]. The mapping generating the iteration is

$$\otimes A(k+1) = G(\otimes A(k)) = f(\otimes W \otimes A) = [f(\otimes w_1 \otimes A), \ldots, f(\otimes w_n \otimes A)],$$

where $\otimes W$ is the weight matrix of grey weights, $\otimes w_j$ is its jth row and f is the sigmoid threshold function with steepness parameter λ. Here fixed point $\otimes A^*$ is

$$\otimes A^* = [\otimes A_1^*, \ldots, \otimes A_n^*]^T \in \left[[\underline{A_1^*}, \overline{A_1^*}], \ldots, [\underline{A_n^*}, \overline{A_n^*}]\right]^T.$$

The grey fixed point is unique in the sense that the endpoints of the intervals containing grey concept values are unique, i.e. the values $\underline{A_i^*}$ and $\overline{A_i^*}$ are unique for every i. The proofs of the theorems are based on Banach's contraction mapping theorem [8]. Roughly speaking, if one can find a c independent of $\otimes A$ and $\otimes A'$, such that $d(G(\otimes A), G(\otimes A') \leq cd(\otimes A, \otimes A')$, then mapping G has one and only one fixed point. Here d is a distance metric defined between the concept vectors. The general properties of a distance metric are the following:

1. $d(x, y) \geq 0$ (non-negativity)
2. $d(x, y) = 0$ if and only if $x = y$. (indescernibility)
3. $d(x, y) = d(y, x)$ (symmetry)
4. $d(x, y) + d(y, z) \geq d(x, z)$ (triangle inequality or subadditivity)

In the proofs of the theorems regarding the unique fixed point of a FGCM, we define and apply different distance metrics.

Lemma 10.1 *Let A and B be vectors of length n, composed of grey numbers Then the equation with a p-norm* $\|x\|_p = \left(\sum_{i=1}^{n} |x_i|^p\right)^{1/2}$, $p \geq 1$

$$d_p(A, B) = \left(\frac{\|\underline{A} - \underline{B}\|_p^p + \|\overline{A} - \overline{B}\|_p^p}{2}\right)^{1/p} \tag{10.8}$$

defines a distance metric between vectors of grey numbers.

Proof To verify that it is a distance metric in mathematical sense, we check the four properties listed above which a distance metric should satisfy.

1. (non-negativity) The inequality $d_p(A, B) \geq 0$ trivially holds.
2. (indescernibility) If $A = B$, then their left and right are the same coordinate-wise. It means that $\|\underline{A} - \underline{B}\|_p = 0$ and $\|\overline{A} - \overline{B}\|_p = 0$, thus $d_p(A, B) = 0$.
3. (symmetry) It is again straightfoward.
4. (triangle inequality)

$$\begin{aligned}
d_p(A, C) &= \left(\frac{\|\underline{A} - \underline{C}\|_p^p + \|\overline{A} - \overline{C}\|_p^p}{2}\right)^{1/p} \\
&= \left(\frac{\|\underline{A} - \underline{C} + \underline{B} - \underline{B}\|_p^p + \|\overline{A} - \overline{C} + \overline{B} - \overline{B}\|_p^p}{2}\right)^{1/p} \\
&= \left(\frac{\|\underline{A} - \underline{B} + \underline{B} - \underline{C}\|_p^p + \|\overline{A} - \overline{B} + \overline{B} - \overline{C}\|_p^p}{2}\right)^{1/p} \\
&\leq \left(\frac{\|\underline{A} - \underline{B}\|_p^p + \|\underline{B} - \underline{C}\|_p^p + \|\overline{A} - \overline{B}\|_p^p + \|\overline{B} - \overline{C}\|_p^p}{2}\right)^{1/p} \\
&\leq \left(\frac{\|\underline{A} - \underline{B}\|_p^p + \|\overline{A} - \overline{B}\|_p^p}{2}\right)^{1/p} + \left(\frac{\|\underline{B} - \underline{C}\|_p^p + \|\overline{B} - \overline{C}\|_p^p}{2}\right)^{1/p} \\
&= d_p(A, B) + d_p(B, C)
\end{aligned}$$

Since all of the properties are fulfilled, it is really a distance metric. □

Theorem 10.1 *Let* $\otimes W$ *be the weight matrix (including possible feedback) of a fuzzy grey cognitive map (FGCM), where the weights* $\otimes w_{ij}$ *are nonnegative or nonpositive grey numbers and let* $\lambda > 0$ *be the parameter of the sigmoid function* $f(x) = 1/(1 + e^{-\lambda x})$ *applied for the iteration. Let* w_{ij}^* *be defined as in Eq. 10.7. Moreover, let* W^* *be a matrix defined by the* w_{ij}^* *values. If the inequality*

$$\|W^*\|_p < \frac{4}{\lambda} \tag{10.9}$$

holds for any ≥ 1*, then the FGCM has one and only one grey fixed point, regardless of the initial concept values.*

Proof We are going to show that if the condition stated in the theorem holds, then the mapping generating the iteration is a contraction mapping w.r.t. distance metric $d_p(A, B)$. Consequently, it has exactly one fixed point and it is the limit of the iteration for every initial concept vector.

$$\begin{aligned}
\left|f(\underline{w_i A}) - f(\underline{w_i A'})\right|^p &\leq \left(\frac{\lambda}{4}\right)^p \left|\underline{w_i A} - \underline{w_i A'}\right|^p \\
\left|f(\overline{w_i A}) - f(\overline{w_i A'})\right|^p &\leq \left(\frac{\lambda}{4}\right)^p \left|\overline{w_i A} - \overline{w_i A'}\right|^p
\end{aligned} \tag{10.10}$$

Unfortunatelly, the equations

$$\underline{w_i A} = \sum_{j=1}^{n} \underline{w_{ij}} \cdot \underline{A_j} \tag{10.11}$$

$$\overline{w_i A} = \sum_{j=1}^{n} \overline{w_{ij}} \cdot \overline{A_j} \tag{10.12}$$

hold only in the case when all of weights are nonnegative. In general

$$\otimes w_{ij} A_j \in \begin{cases} \left[\underline{w_{ij}} \cdot \underline{A_j},\ \overline{w_{ij}} \cdot \overline{A_j}\right] \text{ if } \otimes w_{ij} \geq 0 \\ \\ \left[\underline{w_{ij}} \cdot \overline{A_j},\ \overline{w_{ij}} \cdot \underline{A_j}\right] \text{ if } \otimes w_{ij} \leq 0 \end{cases} \tag{10.13}$$

We handle the cases step-by-step, starting with the most straightforward one. In the case when all of the weights are nonnegative we get the following upper estimation:

$$\begin{aligned}
\left|f(\underline{w_i A}) - f(\underline{w_i A'})\right|^p &\leq \left(\frac{\lambda}{4}\right)^p \left|\underline{w_i A} - \underline{w_i A'}\right|^p \\
&= \left(\frac{\lambda}{4}\right)^p \left|\sum_{j=1}^{n} \underline{w_{ij}} \cdot \left(\underline{A_j} - \underline{A'_j}\right)\right|^p
\end{aligned} \tag{10.14}$$

Let us consider the case when all of the weights are nonpositive. The argument is similar to the previous one, with a small modification:

$$\left|f(\underline{w_i A}) - f(\underline{w_i A'})\right|^p \leq \left(\frac{\lambda}{4}\right)^p \left|\underline{w_i A} - \underline{w_i A'}\right|^p$$
$$= \left(\frac{\lambda}{4}\right)^p \left|\sum_{j=1}^{n} \underline{w_{ij}} \cdot \left(\overline{A_j} - \overline{A'_j}\right)\right|^p \tag{10.15}$$

Similar inequalities hold for the right (upper) endpoints, too.

In the general case, when positive and negative weights also occur, in the upper bound of $\left(f(\underline{w_i A}) - f(\underline{w_i A'})\right)^2$ may occur terms like $\underline{w_{ij}} \cdot \left(\underline{A_j} - \underline{A'_j}\right)$ and terms like $\underline{w_{ij}} \cdot \left(\overline{A_j} - \overline{A'_j}\right)$. The following facts come trivially from the two cases we have discussed already:

- If $\underline{w_{ij}} \cdot \left(\underline{A_j} - \underline{A'_j}\right)$ occurs in the upper bound of $\left(f(\underline{w_i A}) - f(\underline{w_i A'})\right)^2$, then $\overline{w_{ij}} \cdot \left(\overline{A_j} - \overline{A'_j}\right)$ occurs in the upper bound of $\left(f(\overline{w_i A} - f(\overline{w_i A'})\right)^2$.
- If $\underline{w_{ij}} \cdot \left(\overline{A_j} - \overline{A'_j}\right)$ occurs in the upper bound of $\left(f(\underline{w_i A}) - f(\underline{w_i A'})\right)^2$, then $\overline{w_{ij}} \cdot \left(\underline{A_j} - \underline{A'_j}\right)$ occurs in the upper bound of $\left(f(\overline{w_i A}) - f(\overline{w_i A'})\right)^2$.

The following inequalities come from the definition of w^*_{ij}:

$$\left|\underline{w_{ij}} \cdot \left(\underline{A_j} - \underline{A'_j}\right)\right| \leq w^*_{ij} \cdot \left|\underline{A_j} - \underline{A'_j}\right| \tag{10.16}$$
$$\left|\overline{w_{ij}} \cdot \left(\overline{A_j} - \overline{A'_j}\right)\right| \leq w^*_{ij} \cdot \left|\overline{A_j} - \overline{A'_j}\right| \tag{10.17}$$
$$\left|\underline{w_{ij}} \cdot \left(\overline{A_j} - \overline{A'_j}\right)\right| \leq w^*_{ij} \cdot \left|\overline{A_j} - \overline{A'_j}\right| \tag{10.18}$$
$$\left|\overline{w_{ij}} \cdot \left(\underline{A_j} - \underline{A'_j}\right)\right| \leq w^*_{ij} \cdot \left|\underline{A_j} - \underline{A'_j}\right| \tag{10.19}$$

Applying these upper estimations and rearranging the terms in the the upper estimation of $d^p(\otimes G(A), \otimes G(A'))$ by $\left|\underline{A_j} - \underline{A'_j}\right|^p$ and $\left|\overline{A_j} - \overline{A'_j}\right|^p$ we get that

$$d(\otimes G(A), \otimes G(A')) \leq \frac{\lambda}{4}\left[\frac{1}{2}\left(\sum_{i=1}^{n} |\underline{w_i A} - \underline{w_i A'}|^p + \sum_{i=1}^{n} |\overline{w_i A} - \overline{w_i A'}|^p\right)\right]^{1/p} \tag{10.20}$$
$$\leq \frac{\lambda}{4}\left[\frac{1}{2}\left(\sum_{i=1}^{n}\sum_{j=1}^{n} {w^*_{ij}}^p |\underline{A} - \underline{A'}|^p + \sum_{i=1}^{n}\sum_{j=1}^{n} {w^*_{ij}}^p |\overline{A} - \overline{A'}|^p\right)\right] \tag{10.21}$$
$$= \frac{\lambda}{4}\left[\frac{1}{2}\left(\|W^*(\underline{A} - \underline{A'})\|_p^p + \|W^*(\overline{A} - \overline{A'})\|_p^p\right)\right]^{1/p} \tag{10.22}$$
$$\leq \frac{\lambda}{4}\left[\frac{1}{2}\left(\|W^*\|_p^p \|\underline{A} - \underline{A'}\|_p^p + \|W^*\|_p^p \|\overline{A} - \overline{A'}\|_p^p\right)\right]^{1/p} \tag{10.23}$$

$$= \frac{\lambda}{4}\|W^*\|_p \left[\frac{\|\underline{A} - \underline{A'}\|_p^p + \|\overline{A} - \overline{A'}\|_p^p}{2}\right]^{1/p} \tag{10.24}$$

$$= \frac{\lambda}{4}\|W^*\|_p d(A, A') \tag{10.25}$$

According to Banach's fixed point theorem, if $\frac{\lambda}{4}\|W^*\|_p < 1$, then mapping G is a contraction, so it has one and only one fixed point, which completes the proof. □

The $p = \infty$ (infinity norm) case can be handled as the limit of the previous case when $p \to \infty$. Nevertheless, we state this separately, since taking the limit is not an obvious step in applications, what we need is short formula.

Theorem 10.2 *Let $\otimes W$ be the weight matrix (including possible feedback) of a fuzzy grey cognitive map (FGCM), where the weights $\otimes w_{ij}$ are nonnegative or nonpositive grey numbers and let $\lambda > 0$ be the parameter of the sigmoid function $f(x) = 1/(1 + e^{-\lambda x})$ applied for the iteration. Let w_{ij}^* be defined as in Eq. 10.7. Moreover, let W^* be a matrix defined by the w_{ij}^* values. If the inequality*

$$\|W^*\|_\infty < \frac{4}{\lambda} \tag{10.26}$$

holds, then the FGCM has one and only one grey fixed point, regardless of the initial concept values.

Proof Let us define the distance of grey concept vectors A and A' as

$$d(\otimes A, \otimes A') = \max_i \max\left\{\left|\underline{A_i} - \underline{A_i'}\right|, \left|\overline{A_i} - \overline{A_i'}\right|\right\} \tag{10.27}$$

In the proof, we apply the contraction mapping theorem using this distance metric.

$$\begin{aligned} d(\otimes G(A), \otimes G(A')) &= \max_i \max\left\{\left|\underline{G(A)_i} - \underline{G(A')_i}\right|, \left|\overline{G(A)_i} - \overline{G(A')_i}\right|\right\} \\ &\leq \max_i \max\left\{\frac{\lambda}{4}\left|\underline{w_i A} - \underline{w_i A'}\right|, \frac{\lambda}{4}\left|\overline{w_i A} - \overline{w_i A'}\right|\right\} \end{aligned} \tag{10.28}$$

For further upper estimation, we use the following inequalities:

$$\sum_{j=1}^n w_{ij}^*\left|\underline{A_j} - \underline{A_j'}\right| \leq \sum_{j=1}^n w_{ij}^* \max_j \left|\underline{A_j} - \underline{A_j'}\right| \tag{10.29}$$

$$\sum_{j=1}^n w_{ij}^*\left|\overline{A_j} - \overline{A_j'}\right| \leq \sum_{j=1}^n w_{ij}^* \max_j \left|\overline{A_j} - \overline{A_j'}\right| \tag{10.30}$$

So we get that

$$\max\left\{\left|\underline{w_i A} - \underline{w_i A'}\right|, \left|\overline{w_i A} - \overline{w_i A'}\right|\right\} \leq \sum_{j=1}^{n} w_{ij}^* \max\left\{\left|\underline{A_j} - \underline{A'_j}\right|, \left|\overline{A_j} - \overline{A'_j}\right|\right\} \tag{10.31}$$

Putting these altogether:

$$\begin{aligned} d(\otimes G(A), \otimes G(A')) &\leq \frac{\lambda}{4} \max_i \max\left\{\left|\underline{w_i A} - \underline{w_i A'}\right|, \left|\overline{w_i A} - \overline{w_i A'}\right|\right\} \\ &\leq \frac{\lambda}{4} \max_i \left\{\sum_{j=1}^{n} w_{ij}^* \max\left\{\left|\underline{A_j} - \underline{A'_j}\right|, \left|\overline{A_j} - \overline{A'_j}\right|\right\}\right\} \\ &\leq \frac{\lambda}{4} \max_i \left\{\sum_{j=1}^{n} w_{ij}^*\right\} \max_j \max\left\{\left|\underline{A_j} - \underline{A'_j}\right|, \left|\overline{A_j} - \overline{A'_j}\right|\right\} \end{aligned} \tag{10.32}$$

From the fact, that $w_{ij}^* \geq 0$ and from the definition of the distance applied we get that:

$$\|W^*\|_\infty = \max_i \left\{\sum_{j=1}^{n} w_{ij}^*\right\} \tag{10.33}$$

$$d(\otimes A, \otimes A') = \max_j \max\left\{\left|\underline{A_j} - \underline{A'_j}\right|, \left|\overline{A_j} - \overline{A'_j}\right|\right\} \tag{10.34}$$

Consequently, the final form of the last inequality is the following:

$$d(\otimes G(A), \otimes G(A')) \leq \frac{\lambda}{4} \|W^*\|_\infty d(\otimes A, \otimes A') \tag{10.35}$$

If the coefficient of the distance between A and A' is less than one, then this mapping is a contraction, consequently it has exactly one fixed point. This is equivalent to the statement that $\|W^*\|_\infty < 4/\lambda$, which was the condition in the theorem. □

10.3 Fuzzy Cognitive Maps with Fuzzy Weights

There are many systems, where not only the parameters, but the causal effect are not known exactly. Typical examples are large systems involving human decisions, public opinion etc., for instance waste management, urban planning, risk analysis or business decisions. In these cases almost impossible to describe the causal connection by an exact value, the experts rather prefer terms like *strong*, *medium*, *weak* etc.

The iteration rule applied in fuzzy cognitive maps requires computations with these values, that can be represented by fuzzy sets [1, 2, 10]. First, we review the most basic mathematical tools for the extension of real operations to fuzzy sets, then we prove a theorem regarding the unique fixed point of a fuzzy set valued cognitive map.

10.3.1 Some Basic Notions of Fuzzy Mathematics

The approach is based on the α-cut-wise operations between fuzzy numbers. We shortly summarize the most important notions and theorems applied in the mathematical explanation and derivation of the results.

Fuzzy numbers are convex, normal fuzzy sets with bounded support.

Definition 10.1 The α-cut of a fuzzy number A is a non-fuzzy set denoted by $[A]^\alpha$ and defined by

$$[A]^\alpha = \begin{cases} \{x \in X \mid A(x) \geq \alpha\} & \text{if } \alpha > 0 \\ \mathrm{cl}(\mathrm{supp} A) & \text{if } \alpha = 0 \end{cases} \tag{10.36}$$

where $\mathrm{cl}(\mathrm{supp} A)$ denotes the closure of the support of A.

The α-cuts of a fuzzy number are intervals (because of fuzzy convexity), so they can be determined by the left and right endpoints:

$$[A]^\alpha = \left[[A]^\alpha_L,\ [A]^\alpha_R\right] \tag{10.37}$$

or we may use other notation:

$$[A]^\alpha = \left[\underline{A}_\alpha,\ \overline{A}_\alpha\right] \tag{10.38}$$

The most well-known extension of real functions to fuzzy numbers was given by Zadeh.

Definition 10.2 ([11]) Let $f\colon X_1 \times X_2 \times \ldots \times X_n \to Y$ be a mapping. Let A_1, $A_2, \ldots, A_n$ be fuzzy subsets of $X_1, X_2, \ldots, X_n$, respectively. Then $f(A_1, A_2, \ldots, A_n) = B$, where B is a fuzzy subset of Y such that

$$B(y) = \sup\left\{\min\{A_1(x_1), \ldots, A_n(x_n)\} \mid (x_1, \ldots, x_n) \in f^{-1}(y)\right\} \tag{10.39}$$

if if $f^{-1}(y) \neq \emptyset$, and 0 otherwise.

In this case the α-cuts of membership function of f:

Theorem 10.3 ([7]) *Let $f\colon X \times X \times \ldots \times X \to X$ be a continuous function and let $A_1, A_2, \ldots, A_n$ be fuzzy numbers. Then*

$$[f(A_1, A_2, \ldots, A_n)]^\alpha = f([A_1]^\alpha, [A_2]^\alpha, \ldots, [A_n]^\alpha) \tag{10.40}$$

where $f(A_1, A_2, \dots, A_n)$ *is defined by the extension principle and*

$$f([A_1]^\alpha, [A_2]^\alpha, \dots, [A_n]^\alpha) = \{f(x_1, x_2, \dots, x_n) \mid x_i \in [A_i]^\alpha\}. \tag{10.41}$$

10.3.2 Convergence of Fuzzy Set Valued Cognitive Maps

It is clear from Nguyen's theorem, that the basic arithmetic operations defined by the extension principle are the same as the operations introduced in the previous section regarding grey numbers (intervals). Namely, the α-cut-wise basic arithmetic operations follow the same rules as the well-known interval arithmetic used in the context of grey numbers. Thus, the theorems regarding *intervals* (grey numbers) remain valid α-cut-wise.

Let $\alpha \in [0, 1]$ be given, the connections between the concepts are described by fuzzy numbers w_{ij}. The α-cuts of these fuzzy numbers are intervals with left and right endpoints $\underline{w_{ij}}_\alpha$ and $\overline{w_{ij}}_\alpha$, respectively. Let

$$w^*_{ij_\alpha} = \max\left\{|\underline{w_{ij}}_\alpha|, |\overline{w_{ij}}_\alpha|\right\} \tag{10.42}$$

and W^*_α is the matrix composed of the $w^*_{ij_\alpha}$ values. Then the straightforward mathematical consequence of Theorem 10.1.

Theorem 10.4 *Consider a fuzzy cognitive map with fuzzy weights as described above. Let* $\lambda > 0$ *be the parameter of the sigmoid threshold function. If the inequality*

$$\|W^*_\alpha\|_p < \frac{4}{\lambda} \tag{10.43}$$

holds, then the FCM has a unique fixed point at this α*-cut level.*

Although this theorem is valid, it is not natural to handle the convergence α-cut-wise. Instead, we provide a theorem not only for α-cuts, but for the whole fuzzy set valued FCM. It is based on a kind of aggregation of distance measures defined between the fuzzy concept values. Remember, that the distance between interval-valued concept vectors A and B was defined as

$$d_p(A, B) = \left(\frac{\|\underline{A} - \underline{B}\|_p^p + \|\overline{A} - \overline{B}\|_p^p}{2}\right)^{1/p}.$$

Moreover, we have seen that it is a distance metric in the strict mathematical sense. Using this distance metric, we can define the α-cut-wise distance between fuzzy set valued concept vectors in a similar way. Namely, for a given $\alpha \in [0, 1]$, the distance at level α is just the distance of the vectors with interval (α-cut) components, let us denote it by $d_{p,\alpha}(A, B)$. Then the whole distance of fuzzy sets valued vectors A and B is the following:

$$d_p(A, B) = \left[\int_0^1 d^p_{p,\alpha}(A, B) d\alpha \right]^{1/p} \tag{10.44}$$

As a first step, we show that it is a distance metric.

Lemma 10.2 *Equation 10.44 defines a distance metric between vectors of fuzzy sets.*

Proof We check the required mathematical properties of a distance metric.

1. $d_p(A, B) \geq 0$: it is clear from the definition.
2. $d_p(A, B) = d_p(B, A)$: it also straightforward from the definition.
3. $d_p(A, B) = 0$ if and only if $A = B$:
   If $A = B$, then it is straithforward that $d_p(A, B) = 0$, since we integrate zero. If $d_p(A, B) = 0$, then $d_{p,\alpha}(A, B) \neq 0$ holds only for a countable set of α values. If we assume continuous membership functions, then $d_p(A, B) = 0$ implies $A = B$.
4. Triangle inequality:
   We use the fact that the triangle inequality holds at every α-cut.

$$d^p_p(A, C) = \int_0^1 d^p_{p,\alpha}(A, C) d\alpha \leq \int_0^1 d^p_{p,\alpha}(A, B) + d^p_{p,\alpha}(B, C) d\alpha \tag{10.45}$$

$$= \int_0^1 d^p_{p,\alpha}(A, B) d\alpha + \int_0^1 d^p_{p,\alpha}(B, C) d\alpha = d^p_p(A, B) + d^p_p(B, C) \tag{10.46}$$

Thus,

$$d_p(A, C) \leq d_p(A, B) + d_p(B, C). \tag{10.47}$$

Since all of the four properties are fulfilled, it is a distance metric. □

Then, using this distance metric, we show that under some conditions the mapping G generating the iteration is a *contraction*, so according to Banach's fixed point theorem, it has a unique fixed point (vector of fuzzy sets) and the iteration, regardless of the initial concept vector, finally leads to this point.

Theorem 10.5 *Consider a fuzzy cognitive map, where the weights w_{ij} are either nonnegative or nonpositive fuzzy numbers, so their support is subset of* $[-1, 0]$ *or* $[0, 1]$*. Let w^*_{ij} be the absolute value of the most extreme value of the fuzzy set w_{ij}, i.e.* $w^*_{ij} = \max \left\{ |\underline{w_{ij}}_0|, |\overline{w_{ij}}_0| \right\}$ *(i.e. it is the maximal absolute value of the endpoints of the zero α-cut) and let W^* be the matrix composed of the w^*_{ij} values. Let $\lambda > 0$ be the parameter of the sigmoid function $f(x) = 1/(1 + e^{-\lambda x})$ applied for the iteration. If the inequality*

$$\| W^* \|_p < \frac{4}{\lambda} \tag{10.48}$$

holds with an induced matrix norm, then the FCM has one and only one fixed point, i.e. every initial stimulus leads to the same vector of fuzzy sets.

Proof The α-cuts are intervals and we have seen that the following inequality holds for interval-valued (i.e. grey) vectors, where $[*]^\alpha$ denotes the α-cut, W_α is the matrix of the α-cuts of the weights (matrix of intervals) and W_α^* is a matrix of elements $w_{ij_\alpha}^*$s:

$$d_{p,\alpha}([G(A)]^\alpha, [G(A')]^\alpha) \leq \frac{\lambda}{4} \| W_\alpha^* \|_p d_{p,\alpha} d([A]^\alpha, [A']^\alpha)$$

Consequently, the following inequality holds:

$$\int_0^1 d_{p,\alpha}^p([G(A)]^\alpha, [G(A')]^\alpha) d\alpha \leq \left(\frac{\lambda}{4}\right)^p \int_0^1 \| W_\alpha^* \|_p^p d_{p,\alpha}^p([A]^\alpha, [A']^\alpha) d\alpha.$$

Since the weights (w_{ij}s) are convex fuzzy sets, it follows that $w_{ij_\alpha}^* \leq w_{ij}^*$. Then, from the monotonicity of p-norm, it follows that $\| W_\alpha^* \|_p$ reaches its maximum at $\alpha = 0$, so $\| W_\alpha^* \|_p \leq \| W^* \|_p$. It means that

$$\int_0^1 d_{p,\alpha}^p([G(A)]^\alpha, [G(A')]^\alpha) d\alpha \leq \left(\frac{\lambda}{4}\right)^p \| W^* \|_p^p \int_0^1 d_{p,\alpha}^p([A]^\alpha, [A']^\alpha) d\alpha.$$

Thus,

$$\left[\int_0^1 d_{p,\alpha}^p([G(A)]^\alpha, [G(A')]^\alpha) d\alpha \right]^{1/p} \leq \frac{\lambda}{4} \| W^* \|_p \left[\int_0^1 d_{p,\alpha}^p([A]^\alpha, [A']^\alpha) d\alpha \right]^{1/p},$$

i.e.

$$d_p(G(A), G(A')) \leq \frac{\lambda}{4} \| W^* \|_p d_p(A, A').$$

It means that if $\frac{\lambda}{4} \| W^* \|_p < 1$ (i.e. $\| W^* \|_p < \frac{4}{\lambda}$), then the mapping generating the iteration is a contraction, so it has a unique fixed point (vector of fuzzy sets) and the iteration converges to this vector, regardless of the initial concept vector. □

10.4 Summary

The uncertainty of weights in fuzzy cognitive maps can be view in several different ways. In this chapter, we discussed some dynamical aspects of fuzzy grey cognitive maps (FGCMs), where the uncertainty is represented by an interval of possible values. Additionally, the case when the weights are considered as fuzzy numbers was also discussed.

We derived sufficient conditions for the existence and uniqueness of fixed points of fuzzy grey cognitive maps and cognitive maps with fuzzy sets weights. The conditions are similar, since fuzzy sets are collections of nested α-cuts and from the mathematical point of view these are just intervals. When the conditions are met, the cognitive maps have a unique equilibrium point, regardless of the initial concept values. We should note that these are not necessary, but only sufficient conditions. Consequently, these conditions might be upgraded in the future, for example by taking into account the extent of uncertainty or the shape of the membership functions.

Acknowledgements This work was supported by the project 'Developing personalized insurance products using artificial intelligence', no. 2020-1.1.2-PIACI-KFI-2021-00267, National Research, Development and Innovation Office (NKFIH).

References

1. Dodurka MF, Sahin A, Yesil E, Urbas L (2015) Learning of FCMS with causal links represented via fuzzy triangular numbers. In: 2015 IEEE international conference on fuzzy systems (FUZZ-IEEE). IEEE, pp 1–8
2. Frias M, Filiberto Y, Nápoles G, García-Socarrás Y, Vanhoof K, Bello R (2017) Fuzzy cognitive maps reasoning with words based on triangular fuzzy numbers. In: Mexican international conference on artificial intelligence. Springer, pp 197–207
3. Harmati IÁ, Kóczy LT (2018) On the convergence of fuzzy grey cognitive maps. In: Conference on information technology, systems research and computational physics, pp 74–84. Springer
4. Harmati IÁ, Kóczy LT (2019) On the convergence of sigmoidal fuzzy grey cognitive maps. Int J Appl Math Comput Sci 29(3):453–466
5. Lin Y, Liu S (2006) Solving problems with incomplete information: a grey systems approach. Adv Imaging Electron Phys 141:77–174
6. Liu S, Lin Y (2006) Grey information: theory and practical applications. Springer Science & Business Media
7. Nguyen HT (1978) A note on the extension principle for fuzzy sets. J Math Anal Appl 64(2):369–380
8. Rudin W et al (1964) Principles of mathematical analysis, vol 3. McGraw-Hill, New York
9. Salmeron JL (2010) Modelling grey uncertainty with fuzzy grey cognitive maps. Expert Syst Appl 37(12):7581–7588
10. Yesil E, Dodurka MF, Urbas L (2014) Triangular fuzzy number representation of relations in fuzzy cognitive maps. In: 2014 IEEE international conference on fuzzy systems (FUZZ-IEEE). IEEE, pp 1021–1028
11. Zadeh LA (1975) The concept of a linguistic variable and its application to approximate reasoning—II. Inf Sci 8(4):301–357

GPSR Compliance
The European Union's (EU) General Product Safety Regulation (GPSR) is a set of rules that requires consumer products to be safe and our obligations to ensure this.

If you have any concerns about our products, you can contact us on

ProductSafety@springernature.com

In case Publisher is established outside the EU, the EU authorized representative is:

Springer Nature Customer Service Center GmbH
Europaplatz 3
69115 Heidelberg, Germany

www.ingramcontent.com/pod-product-compliance
Ingram Content Group UK Ltd.
Pitfield, Milton Keynes, MK11 3LW, UK
UKHW022210051026
15624UKWH00001B/87
* 9 7 8 3 0 3 1 3 7 9 5 8 1 *